Developments in Petroleum Science
Volume 63

Integrated Sand Management For Effective Hydrocarbon Flow Assurance

Series Editor

John Cubitt

Holt, Wales

Developments in Petroleum Science

Volume 63

Integrated Sand Management For Effective Hydrocarbon Flow Assurance

Babs Oyeneyin

Professor of Petroleum Engineering

Robert Gordon University, Aberdeen, UK and

Intelligent Flow Solutions Ltd

Edinburgh, UK

ELSEVIER

AMSTERDAM • BOSTON • HEIDELBERG • LONDON • NEW YORK • OXFORD

PARIS • SAN DIEGO • SAN FRANCISCO • SINGAPORE • SYDNEY • TOKYO

Elsevier
Radarweg 29, PO Box 211, 1000 AE Amsterdam, Netherlands
The Boulevard, Langford Lane, Kidlington, Oxford OX5 1GB, UK
225 Wyman Street, Waltham, MA 02451, USA

First edition 2015

ISBN: 978-0-444-62637-0
ISSN: 0376-7361

To Bunmi, Funmi, Toyin, and Kemi
For being my source of joy always

Contents

Series Editor's Preface

This is the fourth book in the *Developments in Petroleum Science* series since it incorporated *The Handbook of Petroleum Exploration and Production* in 2013. After books on geophysics, stratigraphic reservoir characterization and petrophysics, we now switch our attention to a major engineering problem in hydrocarbon production – integrated sand management.

In hydrocarbon reservoirs that for example characterize the Tertiary basins offshore major delta complexes such as the Gulf Coast of the USA, the Niger Delta of Nigeria, and the Nile Delta of Egypt, thick sequences of sediment are being deposited resulting in rapid subsidence along the basin margins. This leads frequently to over pressure development which in turn restricts cementation and consolidation of the sands. Consequently, when these reservoirs are put into oil or gas production, there is nothing to stop the sand being produced as well.

Not that sand production is restricted to unconsolidated Tertiary reservoirs. Poorly consolidated or cemented sandstones can be prone to sand production under certain circumstances. Mechanical rock failure can in fact be induced by varying factors including the earth's natural stress, inherent strength of the sandstone or more commonly production or drilling activities.

Whatever the origins of sand production, it can be a serious problem in the daily operations of hydrocarbon fields. Sand can cause major damage by eroding and plugging platform or pipeline facilities, by blocking tubulars and separators, or creating cavities behind the pipe. This can ultimately prove to be very expensive if production has to be stopped or surface equipment is damaged. In addition, the produced sand must be separated from the fluids and disposed of at the surface. The challenge then is to keep formation sand in place without unduly restricting production.

This book reviews the causes of sand production and how it can be predicted and controlled. It should therefore be essential reading for those petroleum and reservoir engineers active in production from reservoirs prone to sand production.

John Cubitt
Holt, Wales

Preface

The major challenges for oil and gas industry (OGI) operators in the development of their oil fields are how to:

- Maximise any individual wells being drilled and their performance
- Effectively manage the overall field development operation
- Minimise operational costs and non-productive time and guarantee flow assurance end-to end across the entire composite production system from reservoirs through the wellbore to the topside and flow lines

Flow assurance in the OGI is an engineering process that includes the technology and methods applied to guarantee optimum hydrocarbon delivery from the reservoir through the well to the surface facilities through optimum process and integrated production management that prevents or remediates the problems that inhibit flow, such as sand production/deposition, hydrate nucleation, wax and asphaltene deposition, corrosion, erosion, scaling, emulsions, foaming, and severe slugging.

Sand production is now known to be the biggest problem and one of the major challenges facing the oil/gas industry. In fact, more than 70% of oil/gas reservoirs/fields worldwide are in various degrees of unconsolidation with extensive sanding problems. The majority of these are mature fields characterised by pressure decline and increasing water production that aggravate the sand production.

To increase hydrocarbon supply, the OGI is exploring the emerging frontiers of the deepwater and arctic environments that form the cornerstone of many future oilfield developments[1]. In these environments and also in mature fields, multiphase fluid (oil, gas, water, and solids) production is inevitable and the challenges of multiphase production are in the multiphase separation and monitoring of production. They are usually dominated by varying pressure/temperature conditions in unconsolidated reservoir rocks with the potential for rapid depressurisation, sand production, early water breakthrough, and major multiphase fluid production flow assurance issues. In the case of the mature and younger hydrocarbon-bearing sandstone rocks in a deepwater/arctic environment, sand production is one of the biggest problems for the oil/gas industry.

For deepwater assets, subsea development with long subsea tiebacks to process facilities is now considered to be the cheapest and most cost-effective field development option, but it requires a sand management strategy. This is especially true in the Niger Delta as activities extend to the more challenging deepwater environments of offshore West Africa.

Debate on the most effective sand management strategy has been ongoing many years, with different companies adopting different strategies. Development of an appropriate strategy for the management of reservoirs with sanding problems is rather complex and requires an integrated approach to finding the optimum solution. It requires integration of key aspects of reservoir characterisation, drilling, completion, and production technologies, including sand tolerances (seabed wellhead/flowlines, topside facilities). Mitigation of sanding requires reliable sanding prediction, precise well design, accurate technology selection as well as an optimum completion strategy. These are transient phenomena requiring a life-cycle real-time management.

For the new breed of sand management engineers there is a need to develop a high level of competency to enable them to proffer solutions that will enable them to manage the sand optimally with no loss of productivity as well as minimise the impact of sand production. The key goals for total sand management include:

- Knowing if and when the field or development wells will produce sand
- Understanding how much sand will be produced
- Sand monitoring
- Management of sand production
- Choosing downhole or surface sand control
- Sand transport analysis

This book focuses on addressing these key issues.

Chapter 1 covers:

- Overview of Global Energy and Hydrocarbon Development
- Examples of Deepwater Developments of the World
- Introduction to Deepwater Business Drivers and Deepwater Key Technologies
- Offshore Development Options

Chapters 2 and 3 present:

- Introduction to the Composite Hydrocarbon Production Systems and Subsea System Options and Operations
- Flow Assurance Issues and Fundamental Principles of Management of Reservoirs with Sanding Problems

Chapters 4 and 5 address:

- Fundamentals of Petrophysics and Geomechanical Aspects of Sand Production Forecasts for Deepwater and HP-HT Environments
- Introduction to Sand and Condition Monitoring Strategies for Asset Integrity

Chapters 6, 7, and 8 cover:

- Surface and Subsurface Sand Control Completion Strategy
- Multiphase Sand Transport in Pipelines/Subsea Tiebacks.
- Risk Assessment Criteria for Effective Sand Management and Flow Assurance Management
- Performance Criteria for Reservoirs with Sanding Problems
- Life-cycle Hydrocarbon Asset Management for Flow Assurance

Babs Oyeneyin

Acknowledgements

The author wishes to express his sincere appreciation and gratitude to Elsevier for the opportunity to publish this book and especially to Susan Dennis (Acquisition Editor), Derek Coleman (Editorial Project Manager), and John Cubitt (Developments in Petroleum Science Series Editor) for the motivation, support, and immense contribution to the review and editing of this book.

The author gratefully acknowledges the permission of Robert Gordon University and Intelligent Flow Solutions to utilise relevant referenced materials in this book. The contributions of the Well Engineering Research Group especially to the integrated sand management research activities at RGU inform some of the contents of this book.

Finally, I am indebted to my family for their understanding and support during the writing of this book.

Babs Oyeneyin

Chapter 1

Introduction to Deepwater Field Development Strategies

Chapter Outline

1.1 INTRODUCTION TO GLOBAL ENERGY AND HYDROCARBON DEVELOPMENT

Petroleum is a complex mixture of hydrocarbons that can occur in liquid and gaseous forms within the pore spaces of conventional reservoir rock or unconventional shale matrix at shallow or great depths on land or in an offshore environment, depending on composition, impurities, and prevailing conditions of pressure and temperature.

The process of hydrocarbon exploration and production starts with a search for the presence of hydrocarbons within the rock matrix, drilling of an appropriate conduit from the surface through various rock types to the depth of interest where the hydrocarbon is 'stored', installation of subsurface and surface process facilities and eventual production, and processing and sale.

The operation is capital intensive and needs to be carried out efficiently and safely with little or no impact on the environment. Thus, the operator must possess the ability to extract the hydrocarbon fluids efficiently and economically from the reservoir rock through the wellbore to the topside production facility all the way to the beach via the pipeline (tieback) over the life of a field within any environment. For land locations, the pipeline tieback is normally in the form of a pipeline from a process facility to the export terminal. In offshore/

Developments in Petroleum Science, Vol. 63. http://dx.doi.org/10.1016/B978-0-444-62637-0.00001-4

deepwater environments the subsea tiebacks connecting all deepwater wellhead manifolds to the process facility are now recognised as one of the cheapest ways to develop in deepwater environments.[1]

Therefore, the objectives of oil company operations can be classified into the following key areas:

(a) Safety, Health, and Environment and Energy Supply Security

(b) Maximisation of Capital Outlay

One major objective of any enterprise is to maximise cash flow and recoverable reserves through:

- Maximisation of production rate
- Maximisation of recovery
- Minimisation of downtime through effective prevention and control of operational problems

(c) Minimisation of Costs

Another key objective is the minimisation of overall costs in an attempt to maximise profit through:

- Minimisation of capital cost: Ensuring that optimum capital expenditure (CAPEX) is maintained to ensure efficient production with minimum downtime
- Minimisation of operating expenditure (OPEX) including:
  - Minimum production cost – Today, the emphasis is on minimising lifting cost per volume (per barrel or m^3) of produced fluids, thereby increasing production of fluid
  - Minimum treatment and workover costs

To achieve these key objectives, today's trend is to set up an integrated project team of different specialists comprising the geology and geophysics team, production geologist, reservoir engineer, drilling engineer, production technologist, etc., who will be responsible for effective reservoir management through strategic planning and optimum well design. The integrated team, and most especially the production technologist, must therefore be very conservative with the different facets of development and operation of the well including:

- Drilling (casing design, drilling/completion fluids selection)
- Completion (design/installation of completion string)
- Production (monitoring well and completion performance)
- Workover/Recompletion (diagnosis/installation of new or improved production system)
- Eventual abandonment (planning depletion profile and identifying candidates and procedure for abandonment)

Today, energy security is a serious challenge with the global demand for hydrocarbons (oil and gas) outweighing other sources of energy such as coal, nuclear, and renewable energies (Figure 1.1).

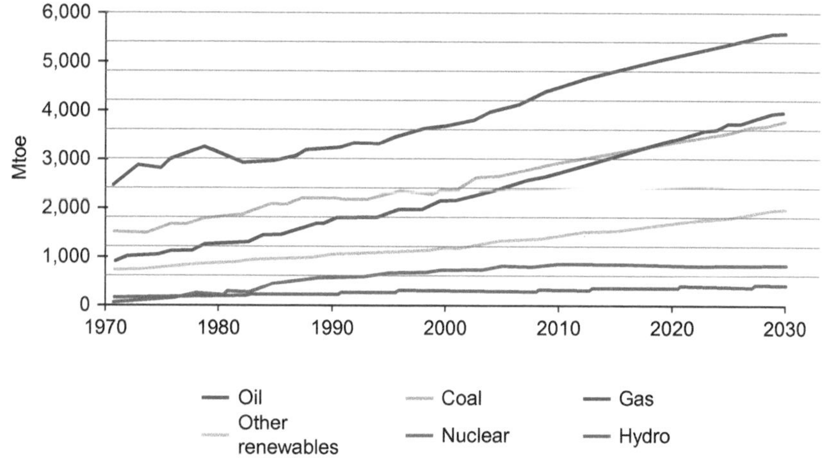

FIGURE 1.1 Global Energy Demand *(Source: International Energy Agency (IEA)[2]).*

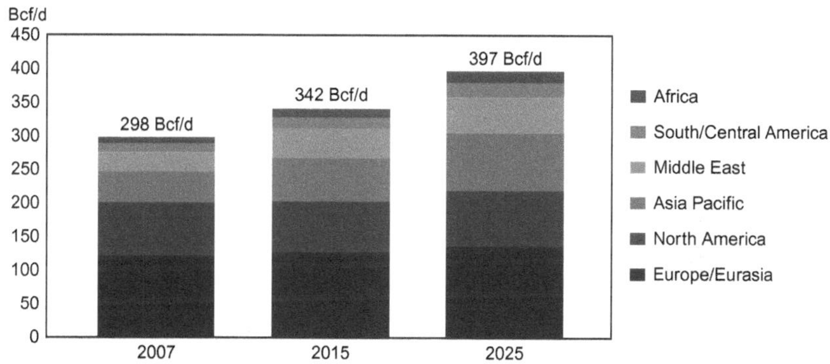

FIGURE 1.2 Global Natural Gas Demand *(Source: International Energy Outlook, IEA[2]).*

The demand for gas (Figure 1.2) is growing faster than the demand for oil (Figure 1.3), especially in developing countries (Figure 1.4).

To meet the ever-increasing gap between the demand and supply of hydrocarbons – especially oil and gas – the petroleum industry is sourcing advanced technologies to enhance production and recovery from the existing mature and marginal fields with more than 50% of recoverable reserves of conventional oil and gas still in place, and carrying out aggressive exploration to new frontiers such as the ultra deepwater environments and arctic regions. Additionally, there is also an increasing trend toward the exploitation of unconventional hydrocarbon reservoirs including shale oil and especially shale gas as well as coal-bed methane and heavy oil. The deepwater environment (from 2000 m to over 3000 m), which includes major assets in the Gulf of Guinea in the West African subregion, the Campos Basin in South America, and the artic regions, now

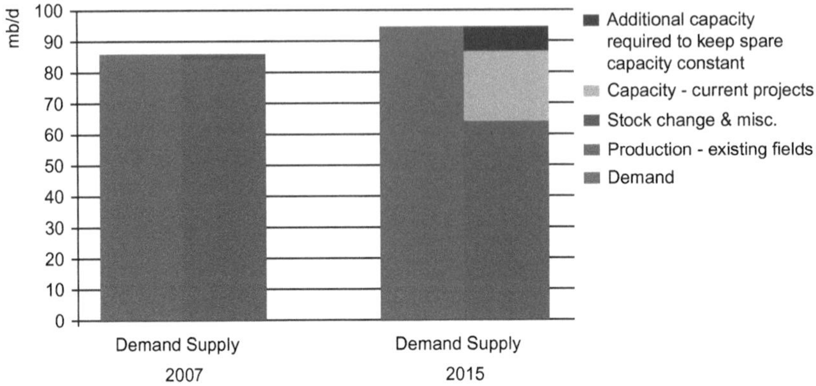

FIGURE 1.3 Oil Production Capacity and Demand Growth to 2015 *(Source: IEA World Energy Outlook 2008[1])*.

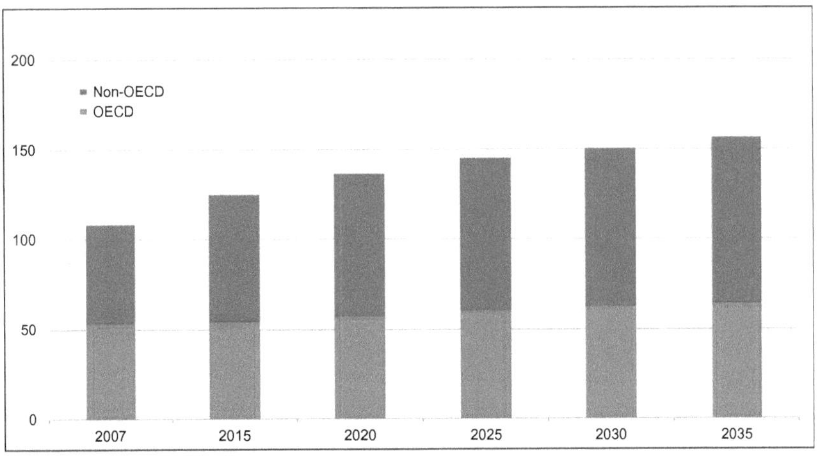

FIGURE 1.4 World Natural Gas Consumption Forecast to 2035.

forms the cornerstone of many oil/gas field developments. These deepwater environments are characterised by very hostile environmental sea conditions and complex temperature and geological features such as high pressure and high temperature (HP-HT) with a very narrow margin of rock pore pressure-fracture pressure window (which makes drilling by conventional overbalanced methods difficult), shallow unconsolidated reservoir sands with massive sand production problems, and a combination of challenging and 'flexible' rock lithologies.

The economic viability of the mature and marginal fields, deepwater and ultra deepwater environments, and even the new frontier of the arctic regions as well as the unconventional shale reservoirs requires the depletion of these reservoirs with a minimum number of wells. To achieve this, engineers are being challenged to use more long horizontal, extended reach, and multilateral wells and long subsea tiebacks. In addition to trying to minimise problems associated with fluid production, there is the added benefit of using these wells to potentially reduce field development costs. In deepwater environments and arctic frontier fields as well as in mature fields, multiphase fluid production accompanied by increasing sand production is inevitable.

The major challenges for companies are how to maximise individual wells and field development performance, minimise operational cost and non-productive time, and guarantee flow assurance across the composite production system from reservoir through wellbore, long subsea tieback to topside for effective asset integrity.

1.2 EXAMPLES OF DEEPWATER DEVELOPMENTS OF THE WORLD

The deepwater environment can be categorised into three main depth tiers:

i. Deep – Tier I
- 500 m – 2000 m of water depth

ii. Very Deep Water Depths – Tier II
- 2000 m – 3000 m of water depth. These represent the current exploration frontiers.

iii. Ultra Deepwater – Tier III
- Greater than 3000 m of water depth

At these water depths, the seabed temperatures can be as low as 100 °F, creating major challenges for flow assurance with cold flow of produced fluid and impact on material integrity.

The major ultra deepwater environments include:

- Arctic
- Gulf of Guinea
- Campos Basin
- Atlantic Margin
- Deepwater Gulf of Mexico
- Caspian
- Australasia

In these environments the lithologies are more complex, with variable pore pressure gradients and a very narrow pore pressure-frac pressure window. They are dominated by very HP-HT unconsolidated reservoirs with the potential for rapid depressurisation, sand production, and very early water breakthrough and major multiphase fluid production issues. Many of the reservoirs are characterised by pressures greater than 20,000 psi and temperatures greater than 350 °F.

FIGURE 1.5 Deepwater Frontiers across the Globe[3].

Popular deepwater locations and activities around the world are presented in Figure 1.5.

1.3 INTRODUCTION TO DEEPWATER BUSINESS DRIVERS

1.3.1 Deepwater Field Development Challenges

Field development challenges can be grouped into the following:

- Capital/Investment Opportunities
- Technical/Operational Challenges
- Research and Development – A number of key areas for technology development essential to the growth of the subsea and deepwater industry include:
    - **i.** Autonomous subsea fields
    - **ii.** Multiphase flow management
    - **iii.** Intelligent autonomous underwater vehicles (AUVs)
    - **iv.** Seabed Processing
    - **v.** Intelligent flow assurance management
    - **vi.** Integrated control system
    - **vii.** New smart materials for SCR (steel catenary riser) and umbilicals
    - **viii.** Wireless sensing
    - **ix.** Cold flow management
- Human Resources – Offshore developments are capital-intensive, requiring specialised equipment and expertise. Skilled manpower for offshore development is in very short supply and the capacity building for the offshore industry is an ongoing challenge.[2]

1.3.2 Deepwater Operational Challenges

Deepwater development operational challenges and business drivers can be divided into four broad categories (Figure 1.6):

1. Depth
2. Complexity of Lithologies
3. Complexity of Reservoir including:
   a. HP-HT reservoirs are defined here as reservoirs with pressures above 3000 psi and temperatures above 300 °F
   b. Unconsolidated sands with high overburden
4. Long-distance pipeline tieback to beach requiring very long subsea tieback and associated solids transport in transient multiphase environment.

For deepwater categories from 500 m and above, it is important to select platforms that are fit for the specific purpose. For 2000 m and above, subsea development and processing often become the cheapest and most reliable field development process, requiring special platforms that are fit-for-purpose and combined with appropriate specialised expertise to support the operation.

At greater depths, the reservoirs, characterised by HP-HT, become more distant from shore and sometimes require the deployment of subsea production systems combined with subsea pipeline tiebacks to transport the produced fluids to platform process facilities or to complete processing subsea and transportation of processed fluids onshore[1,2,4-6].

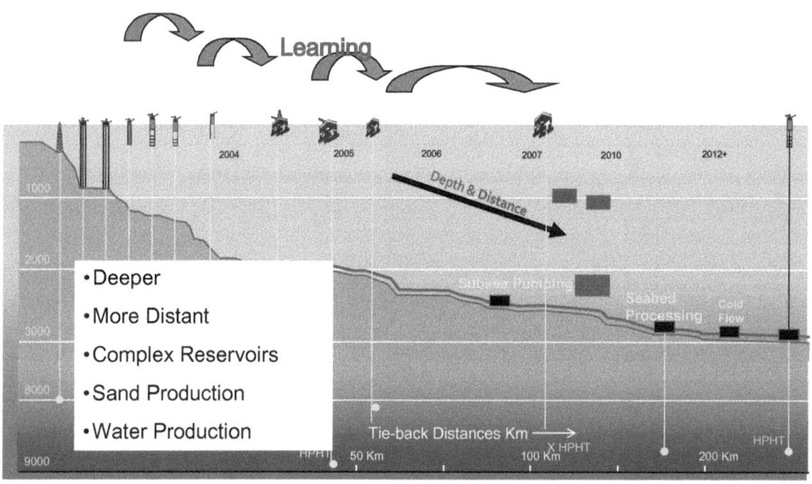

FIGURE 1.6 Deepwater Business Drivers *(Source: Brooks[3])*.

Rigs/platforms in use worldwide vary and include:

- Jackup and concrete platforms
- SPARS
- Drillships
- TLP (tension leg platforms)
- FPSO (floating drilling, production and storage system)
- Semisubmersibles

These systems are much more expensive to manufacture, deploy, and maintain.

1.3.3 Technical Challenges[2–6]

There are serious operational challenges in drilling/completion technology, multiphase production/processing, multiphase transportation of solids arising especially from sand production, and multiphase flow assurance including hydrate, scale, water, and wax/asphaltene formation.

There are also major challenges for deepwater well construction with respect to well integrity, well control, completion strategy, and stability of drilling and completion fluids, subjected to variable HP-HT conditions in the wellbore and high-pressure–low-temperature (HP-LT) in the riser sections.

At water depths of 2000 m and above, subsea production systems with long-distance tiebacks are inevitable with subsea processing and cold flow potentially dominating in the future.[1]

1.3.4 Sand Management Challenges

For mature and younger sandstone reservoirs, especially in deepwater environments, sand production is inevitable. In a deepwater setting, hydrocarbon reservoirs are most often found in turbidite deposits, giant sedimentary systems that form off the mouths of large rivers.

There are also major challenges for deepwater well construction with respect to well integrity, well control, completion strategy, stability of drilling, and completion fluids, subjected to variable HP-HT conditions in the wellbore and HP-LT in the riser sections.

Sand presents a big problem. In many regions, wells can suddenly become inundated by sand, which reduces flow, clogging the bore-hole with dramatic erosion of in-hole components and surface equipment. There is frequently a tight trade-off between the likelihood of sand production and the rate of hydrocarbon extraction. Identification of the cause of issues is difficult, not least because there are so many possibilities, but mis-diagnosis can be astronomically expensive as a consequence.

Every well in such environments has the potential to produce sand now and in the future.

The major challenges for operators in the development of fields with sanding problems are in how to:

- Predict if and when the wells will produce sand, and to quantify the sand production rate.
- Provide input into the subsea and topsides design in terms of possible quantity and particle size distribution of sand, and frequency of sand produced and transported through the wellbore into the subsea and topside facilities
- Optimise the well design so that it is fit-for-purpose.
- Maximise the individual well's production and field development performance
- Effectively manage the sand and multiphase fluid production as well as the facility integrity
- Deploy an appropriate sand management strategy that is fit-for-purpose including critical evaluation of the appropriate sand control method – surface control vs. subsurface control
- Minimise the impact of sand production on wells and subsea production facilities
- Minimise operational cost and non-productive time and guarantee flow assurance end-to-end across the entire composite production system, from reservoirs through the wellbore to the topside and flow lines
- Manage any sand disposal including the environmental impact assessment arising thereof

1.4 WHAT IS INTEGRATED SAND MANAGEMENT?

Integrated sand management can be defined as the consideration of all aspects of subsurface and surface activities related to sand (and other solids) in the efficient production of oil and gas including but not limited to prediction of onset of sand failure and production; surface and subsurface sand control; flow assurance (transportation/erosion); sand monitoring; sand handling; and disposal. It includes the technology deployed and operational procedures/solutions established to minimise the impact of sand production and to improve future well potential/production optimisation including sand transport.

Each of these sand management issues is covered in this book.

Chapter 2

Introduction to the Hydrocarbon Composite Production System

Chapter Outline

Developments in Petroleum Science, Vol. 63. http://dx.doi.org/10.1016/B978-0-444-62637-0.00002-6

2.1 DEFINITION OF PETROLEUM

Petroleum is a complex mixture of hydrocarbons and other constituents that can be in the form of either natural gas or liquid depending on composition, condition of pressure and temperature, reservoir rock depth, and type. The mixture of hydrocarbons is comprised of:

a. Natural gas
b. Crude oil
c. Condensate
d. Other constituants including:
- Nitrogen (N_2)
- Carbon dioxide (CO_2)
- Hydrogen-sulphide H_2S or sulphur

The hydrocarbons comprising the petroleum are largely alkanes; namely:

a. Methane (CH_4)
b. Ethane (C_2H_6)
c. Propane (C_3H_8)
d. n-Butane (n-C_4H_8)
e. Iso-Butane (Iso-C_4H_{10})
f. Pentane (C_5H_{12})
g. Hexane (C_6H_{14})
h. Heptane+(C_7H_{16}+)

2.1.1 Natural Gas

Natural gas is a mixture of individual gas components, predominantly methane, with a number of impurities such as carbon-dioxide, nitrogen, and hydrogen-sulphide. It contains the most volatile hydrocarbon constituents, with $C_1 - C_3$ dominating, and occurs as associated or non-associated gas, is composed of free natural gas in contact with crude oil, and depends on pressure, temperature, and composition of original reservoir fluid.

Natural gas can be classified as:

a. Dry Gas – Natural gas with little or no liquid droplets. Dry gas contains less than two stock bank barrels (STB) of volatilized oil/million ft^3 of gas
b. Wet Gas – Natural gas with appreciable percentage of liquid droplets
c. Sweet Gas – Natural gas with no hydrogen-sulphide
d. Sour Gas – Natural gas with appreciable percentage of hydrogen-sulphide
e. Acid Gas – Natural gas with appreciable percentage of carbon dioxide

These gases are highly compressible, and their behavior can be defined by the perfect gas law and equation of state.

2.1.2 Crude Oil

Crude oil contains liquid hydrocarbons predominately made up of propane and above depending on grade. There are light, medium, or heavy crude oils, depending on their specific gravity (SG) or API (American Petroleum Institute standard) gravity. Crude oils contain many different hydrocarbon compounds characterised into three main families:

1. Paraffins (alkanes)
2. Naphthenes (cycloparaffins)
3. Aromatics

They can be further classified into:

- Volatile oil (SG =0.78-0.825 equivalent to 40°-50° API)
- Black oil (SG=0.825-0.876 equivalent to 30-40°API)
- Heavy oil (SG=0.904-0.934 equivalent to 20°-25°API)
- Tar sands or bitumen (SG=1 equivalent to 10°API)

The light crudes (volatile or black oil) are predominantly paraffin based, while the heavy oil/tar sands are predominantly asphaltene based. Black oil is dominated by a mixture of paraffin and asphaltene.

The API gravity is here defined by

$$^{o}API = \frac{141.5}{SG} - 131.5 \tag{2.1}$$

2.1.3 Condensates

Gas condensates are hydrocarbons in which conditions of temperature and pressure have resulted in the condensation of heavier hydrocarbons from the reservoir gas. Condensates usually occur at a critical pressure and temperature. The API gravity of condensates is of the order of 50°-70° equivalent to SG of 0.702-0.78. An example of a typical reservoir fluid composition is presented in Table 2.1.

2.2 COMPOSITE PRODUCTION SYSTEM

An oil/gas reservoir contains highly compressible fluids at an elevated temperature and pressure. The fluid stores up a considerable energy of compression. Therefore, the efficient production of the fluids requires the effective dissipation of the energy through the entire production system. A typical composite production system for a cased-hole completion with sanding problem (Figure 2.1) is made up of:

- Reservoir (pay zone)
- Perforation

TABLE 2.1 Typical Example of a Range of Hydrocarbon Reservoir Fluid Composition

Component	Dry Gas	Wet Gas	Lean Gas Condensate	Rich Gas Condensate	Near- Critical Gas Conden	Volatile Oil
N_2	0.3	0.4	0.08	2.23	5.65	1.67
CO_2	2.78	0.72	2.44	0.45	2.01	2.18
C1	94.41	81.64	82.10	65.68	46.79	60.51
C2	1.44	8.71	5.78	11.7	12.65	7.52
C3	0.34	4.22	2.87	5.87	5.87	4.74
iC4	0.07	0.68	0.56	1.27	6.04	0
nC4	0.1	1.01	1.23	1.68	-	4.12
iC5	0.04	0.37	0.52	0.71	3.92	0
nC5	0.04	0.29	0.6	0.71	-	2.97
C6	0.04	0.34	0.7	0.998	4.78	1.38
C7+	0.44	1.61	3.12	8.72	12.32	14.91
MW	17	21.4	23.5	35.52	45.05	46.69

MW=Molecular Weight

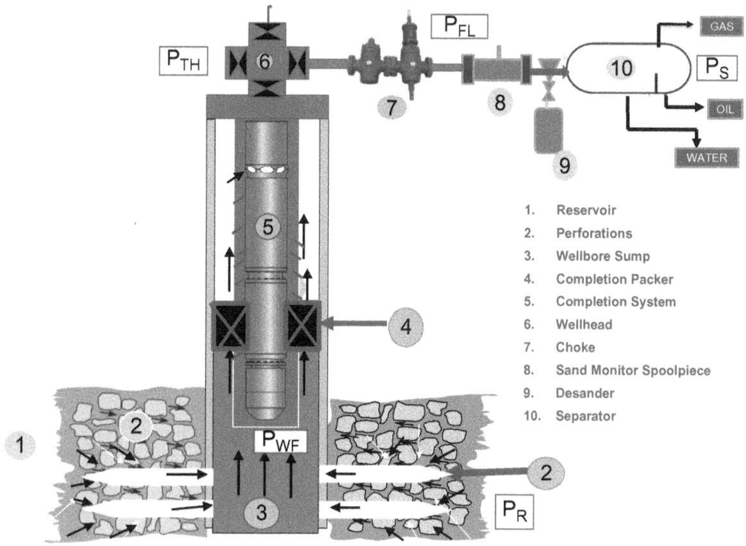

FIGURE 2.1 Composite Production System.

- Wellbore sump
- Production packer
- Production string (completion system)
- Wellhead
- Choke
- Flow line/sand monitoring spool
- Desander
- Separator

The Composite Production System is installed for producing, enhancing, maintaining, and monitoring production with appropriate power supply. First, the reservoir fluid (oil, gas, water, etc.) has to flow through the reservoir rock in order to reach the point of exploitation or production, which is the wellbore. In flowing through the reservoir rock, which is a porous medium, a loss in pressure, otherwise known as reservoir pressure drop or drawdown, will occur. This pressure drop depends on the rock and fluid characteristics.

From the reservoir, the fluid enters the wellbore into the production string (or completion system) with the annulus isolated by the production packer. The completion string is made up principally of production tubing and a combination of different completion component facilities of various sizes installed to perform certain functions. These functions include:

- Fluid circulation
- Isolation & control
- Reservoir monitoring

- Reservoir protection during shutoff
- Flow, pressure, temperature, etc., monitoring
- Well intervention
- Barrier for uncontrolled fluid escape
- Tubing stress tolerance

Flow from the reservoir through perforation tunnels constitutes the wellbore/completion pressure drop. Once inside the wellbore, and flowing up the completion string, it will be subjected to a combination of three principal pressure drops:

(a) Frictional pressure drop, $\Delta P_{Friction}$
(b) Hydrostatic head pressure due to potential energy, P_H
(c) Kinetic energy losses due to expansion/contraction in the fluid flow area and flow through restrictions. ΔP_{KE}

These combinations of pressure drops constitute the vertical lift pressure drop. On arrival at surface, the fluid passes through the surface/wellhead facilities and flow line including passing through the choke. From here, fluid can move via a number of other facilities such as the sand monitoring spool piece and desander unit onto the separator units for the separation into gas, oil, water, etc. These constitute the topside process facilities.

In deepwater environments, the separations can be sub-sea. At depths over 2000 m and above, sub-sea separation can be the most economical way of fluid production management. Alternatively, the use of sub-sea tieback tying in multiple well productions to a single pipeline is an alternative economic option, bringing the produced fluids onshore for further processing. Examples of a typical sub-sea production system are presented in Figures 2.2 and 2.3.

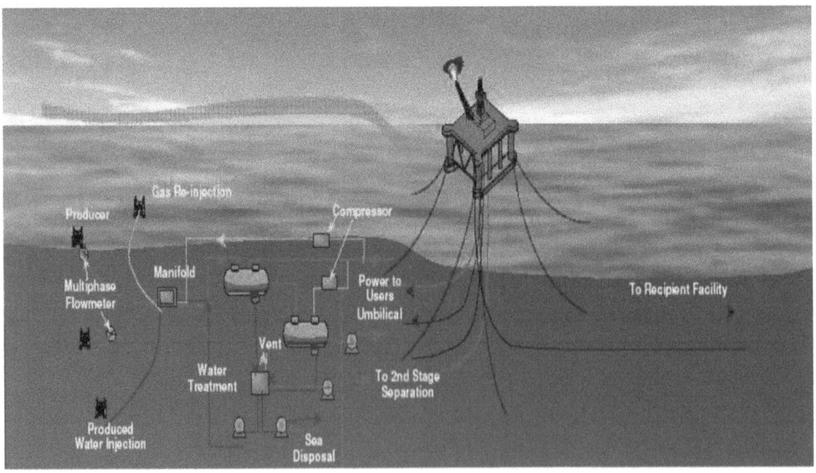

FIGURE 2.2 Typical Sub-sea Production System.

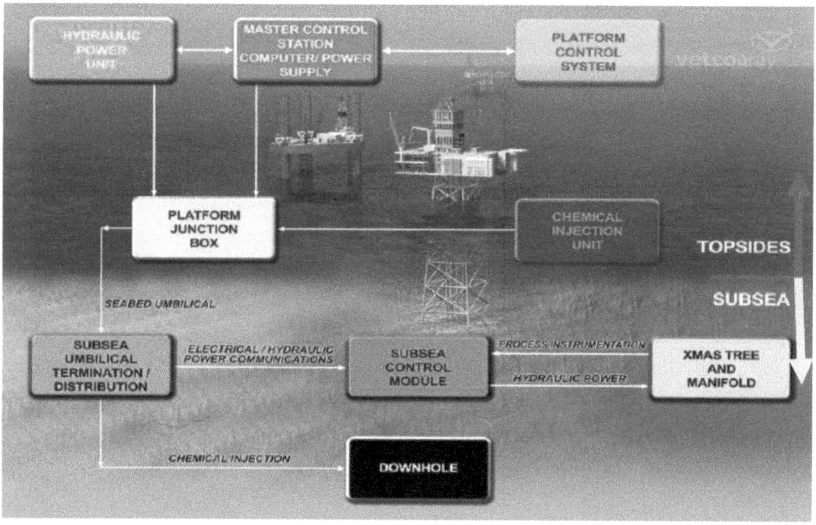

FIGURE 2.3 Schematic of a Typical Deepwater Production System *(Source: Brookes[2]).*

The productivity of the system is dependent on the pressure loss that occurs in each of the units of the system listed. The pressure drop that occurs across the reservoir, the tubing, and choke are production-rate dependent and these relationships define the means by which the production technologist optimises the production of the fluid from the reservoir. Thus, there must be a combined optimisation of:

- Inflow performance (IPR): Definition of pressure-drop versus flow-rate relationship across the reservoir.
- Tubing flow performance (TPR) or wellbore performance (WPR) or vertical lift performance (VLPR): Definition of pressure-drop versus flow-rate relationship in tubing and wellbore necessary to lift the fluid from reservoir to the surface.
- Choke performance (CPR): Definition of pressure drop versus flow relationship involved in choke sizing for flow line pressures to separator.

In the development of a hydrocarbon reservoir, otherwise known as reservoir management, the energy stored within the compressed state of the reservoir fluids in most cases provides the total pressure loss in the entire producing system. Thus,

$$P_{Res} = \Delta P_{Res} + \Delta P_{BHC} + \Delta P_{VL} + \Delta P_{SURF} + \Delta P_{CHOKE} + P_{SEP} \quad (2.2)$$

where:

$\mathbf{P_{Res}}$ = Initial or average pressure within the reservoir

Management Options: Limiting drawdown rate or through fluid injection such as gas, water injection, etc.

$\mathbf{\Delta P_{Res}}$ = Pressure loss caused by the flow of fluid within the reservoir to the wellbore

Management Options: Minimum level of formation damage, and use of more advanced wells such as horizontal or multilateral wells can reduce this. For heavy crudes, reduction of fluid viscosity either by solvent injection or heating can help reduce the reservoir pull. For gas reservoirs, avoiding or limiting the formation of condensate banking will minimise the pressure drop. For condensate reservoirs, avoiding retrograde vaporisation will minimise the reservoir pressure drop.

$\mathbf{\Delta P_{BHC}}$ = Total pressure loss generated by the design of fluid entry into the wellbore, which is dependent on the bottom-hole completion in terms of the sump size

Management Options: The best way to minimise this is to land the production string close to the top of the pay sand and/or use a bridge plug to reduce the sump height.

$\mathbf{\Delta P_{VL}}$ = Pressure loss caused by the fluid flowing up the production string

$$\Delta P_{VL} = \Delta P_{FRICT} + P_H + \Delta P_{KE} \tag{2.3}$$

where:

ΔP_{FRICT} = Friction pressure loss in tubing

P_H = Pressure resistance due to the hydrostatic pressure of the fluid column in the tubing and is dependent on the wellbore fluid density

ΔP_{KE} = Kinetic energy loss due to the velocity head of completion components – fittings, restrictions, subsurface safety valves, etc.

Management Options: From a completion design perspective, a good nodal analysis for optimum string sizing is an essential requirement. The bigger the tubing size the lower the frictional pressure drop and ease of well intervention. For kinetic energy losses, use of inbuilt crossover-subs between joints of different IDs as part completion design helps to reduce acceleration losses. From an operational perspective, minimising scale and wax deposition through the use of inhibitors will help minimise deposition. Use of friction reducers to dope the string during production also helps to minimise frictional losses. For long horizontal and extended reach wells, use of inflow control devices will also help to overcome high frictional drag.

$\mathbf{\Delta P_{SURF}}$ = Pressure loss in wellbore, Xmas tree, and surface flow lines

Management Options: Optimum sizing of flow lines, reducing scale, wax, hydrate, or debris deposition with appropriate inhibitors as well as use of friction reducers are some of the options available. Regular pigging of flow-line and condition monitoring are essential to fluid transportation. Optimisation of booster points can be very relevant to optimising production especially for gas production.

ΔP_{CHOKE} = Pressure loss across the choke designed to cause a significant amount of pressure drop in order to provide stability to downstream separation and treatment facilities

Management Options: Optimisation of choke size in line with desired tubing head flowing pressure is a crucial part of the overall nodal analysis. P_{SEP} is the required operating pressure in separator designed to separate liquid and gas phases from water and base sediments (BSW). The topside facility is an essential part of the composite production system. Separator retention time and removal of base sediments are crucial to the performance of separator units. Fort wells with sanding problems, installation of a hydrocyclone separation system on the platform, or sub-sea wellhead to remove produced sand may be useful in enhancing production. All the pressure drops are production-rate dependent, and the total system pressure drop can be defined as:

$$\Delta P_{TOTAL} = P_{Res} - P_{SEP} = \Delta P_{Res} + \Delta P_{BHC} + \Delta P_{VL} + \Delta P_{SURF} + \Delta P_{CHOKE} \quad (2.4)$$

Thus, each pressure drop can be minimised individually or collectively to produce a maximum production rate for the available pressure. This is known as Production system optimisation.

In general, increased production rate can also be attained through:

1. Increasing the reservoir pressure
2. Providing more energy for the vertical lift process.
3. For reservoirs with sanding problems there is potentially the added benefit of total integrated sand management. The details of how total sand management is achieved are presented in subsequent chapters.

Option 1 is usually more difficult. However, it is possible to ensure effective pressure maintenance through the injection of considerable volume of water or gas and this too can prolong the onset of sand production.

There are several techniques available to improve the vertical lift process. This is known as the artificial lift technique and can include:

(i) Those that simply provide additional energy to assist the lift process. They involve the use of down-hole pumps such as electric submersible pumps (ESPs), hydraulic down-hole pumps, jet pumps, or sucker rod pumps.

(ii) Those that provide some reduction in vertical lift pressure gradient, especially reduction in fluid density. An example of such a technique is a gas lift, which involves the injection of gas into the tubing through the annulus between the production string and casing. Gas mixing with fluid results in the reduction of flowing fluid density. The hydrostatic pressure drops and thus ΔP_{VL} is reduced. This, in addition to the gas expansion, can lead to improved outflow.

Depletion of reservoir energy may finally require the use of secondary/enhanced recovery processes to recover more fluids.

2.3 COMPOSITE PRODUCTION SYSTEM PROFILE

The composite production system is made up of a number of key nodes; namely:

- Node 1 = Reservoir: Inflow performance (IPR) with reservoir pressure, P_R
- Node 2 = Wellbore: Outflow/vertical lift with flowing bottom-hole pressure, P_{wf}
- Node 3 = Production Tubing: Offtake performance
- Node 4 = Well Head/Choke: Choke performance with tubing head pressure, P_{TH}
- Node 5 = Flow line with flow line pressure, P_{FL}
- Node 6 = Separator System: Flow station/topside with separator pressure, P_{SEP}

Optimisation of production and flow efficiency, especially for reservoirs with sanding problems, demands a total systems approach to the management of the overall composite production system. The bottom line is inflow to any node = outflow from the node and pressure exists in each node. Therefore, the systems approach demands robust nodal analysis that depicts node flow rate vs node pressure.

2.3.1 Production System Pressure Profile

A typical node pressures vs flow profile is depicted in Figure 2.4. For wells with surface or down-hole sand control completion the eventual flowing

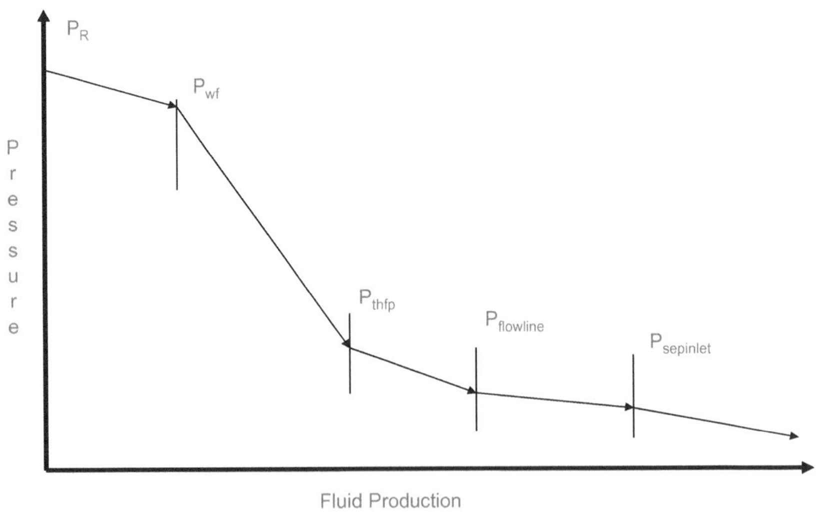

FIGURE 2.4 Typical Production System Pressure vs. Flow Profile.

bottom-hole pressure and corresponding offtake performance will be influenced strongly by the sand control type and system installation/performance.

The system pressure profile is depicted as:

$$P_R - P_{SEP} = \Delta P_{TOTAL} = \Delta P_{RES} + \Delta P_{BHC} + \Delta P_{VL} + \Delta P_{SURF} + \Delta P_{CHOKE} \quad (2.5)$$

where:

- P_R = Reservoir pressure
- P_{SEP} = Separator pressure
- ΔP_{RES} = Drawdown, $P_R - P_{wf}$
- P_{wf} = Flowing bottom-hole pressure
- ΔP_{BHC} = Bottom-hole pressure drop
- ΔP_{VL} = Tubing flow pressure loss, ΔP_{FRIC}+ $\Delta P_{H\ D}$ + ΔP_{KE}
- ΔP_{SURF} = Wellhead, Xmas tree, and surface-line pressure drops
- ΔP_{FRIC} = Frictional pressure drop in production tubing
- ΔP_{KE} = Kinetic energy loss
- P_H = Hydrostatic head, ρ*g*TVD
- TVD = True vertical depth of the well

2.4 RESERVOIR ROCK AND RESERVOIR FLUID PROPERTIES

A typical reservoir rock is made up of a rock matrix, fluids that are a combination of gas, oil, and water, and traps. The focus of this section is on the rock properties and fluids. The hydrocarbon reservoir can be classified in accordance with the following features.

- According to pressure and temperature
  - Normal pressure
  - Abnormal pressure
  - Subnormal
  - Geopressured
- According to lithology
- According to strength
  - Unconsolidated
  - Partially consolidated
  - Friable
  - Consolidated
- According to fluid content
  - Gas reservoir
  - Oil reservoir
  - Condensate reservoir
  - Multiphase reservoir

2.4.1 Pressure and Temperature Regimes

Reservoir rock originally contained formation water (brine), which was subsequently "squeezed" in due to hydrocarbon formation. Therefore, reservoir pressure categorisation is generally defined relative to water density as follows:

- Normal Reservoir: A typical reservoir that supports pressure equivalent to 0.433-0.465 psi/ft
- Geopressured Reservoir: Reservoirs that support pressure > 0.465 psi/ft or greater than 10000 psi
- Subnormal: Pressure < Freshwater weight @ 0.433 psi/ft

Hydrocarbon reservoirs exhibiting:

(i) Pressure > 10,000 psi and Temperature => 300 °F are known as high-pressure high-temperature (HP-HT) reservoirs
(ii) Pressure > 10,000 psi and Temperature =< 60 °F are known as HP-HT reservoirs

HP-HT tiers defined by reservoir temperature and pressure are:

(i) Tier I 10 Kpsi + 350 °F (HP-HT)
(ii) Tier II: 15 Kpsi + 400 °F (Ultra HP-HT)
(iii) Tier IIII: 30 Kpsi + 500 °F (Extreme HP-HT)

These can be found on land locations and in deepwater environments.

Deepwater environments form the cornerstone of many present and future productions and are usually characterised by:

- Complex reservoirs
- HP-HT
- Multiphase fluid production
- Sand production
- Water and multiphase fluid production

Typical deepwater tiers categorised in accordance with water depths are:

- Tier 1: Shallow: 30 m
- Tier II: Deepwater: 30-1500 m
- Tier III: Ultra Deepwater: 1500-3000 m
- Tier IV: Extreme Deepwater: >3000 m

2.4.2 Reservoir Sands

Typical hydrocarbon reservoir sands include carbonates, turbidite/sandstones, and a combination of zechstein and shale barriers. The focus of this book from a sand management perspective is on clastic sandstone reservoirs. Typical

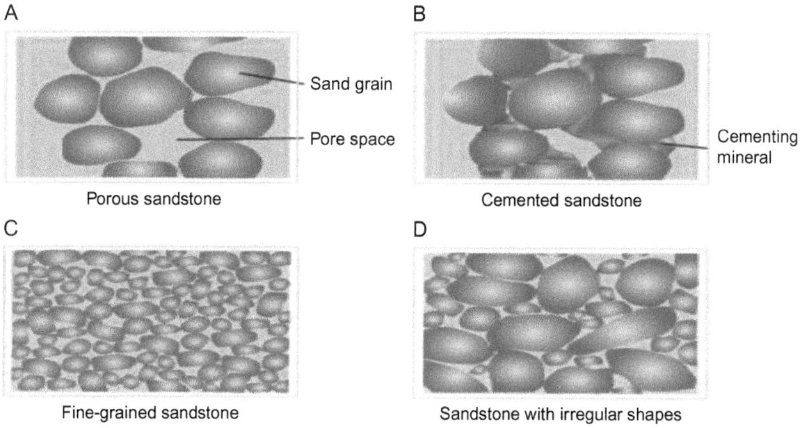

FIGURE 2.5A Typical Sandstone Textural Profiles *(Source: Robert Gordon University (RGU)).*

sandstone packing architectures are presented in Figure 2.4. An example sand matrix (Figure 2.5) is made up of grains, pore space containing the fluid, and grain-grain cementing materials that defines the strength.

2.4.3 Rock Properties

Reservoir rock properties can be categorised into three types of properties:

1. Flow properties
2. Elastic properties – Significantly important to sand production rate prediction
3. Electric properties – Relevant in terms of relationship to well logs and petrophysical properties determination

2.4.3.1 Rock Flow Properties

The key reservoir rock flow properties are:

a. Porosity
b. Permeability
   i. Absolute permeability – Darcy's Law
   ii. Effective and relative permeability
c. Compressibility
d. Wettability
e. Capillary pressure
f. Diffusivity constant
g. Rock textural properties (grain size, pore size distribution, sorting, uniformity coefficient, shape, pack structure coefficient)

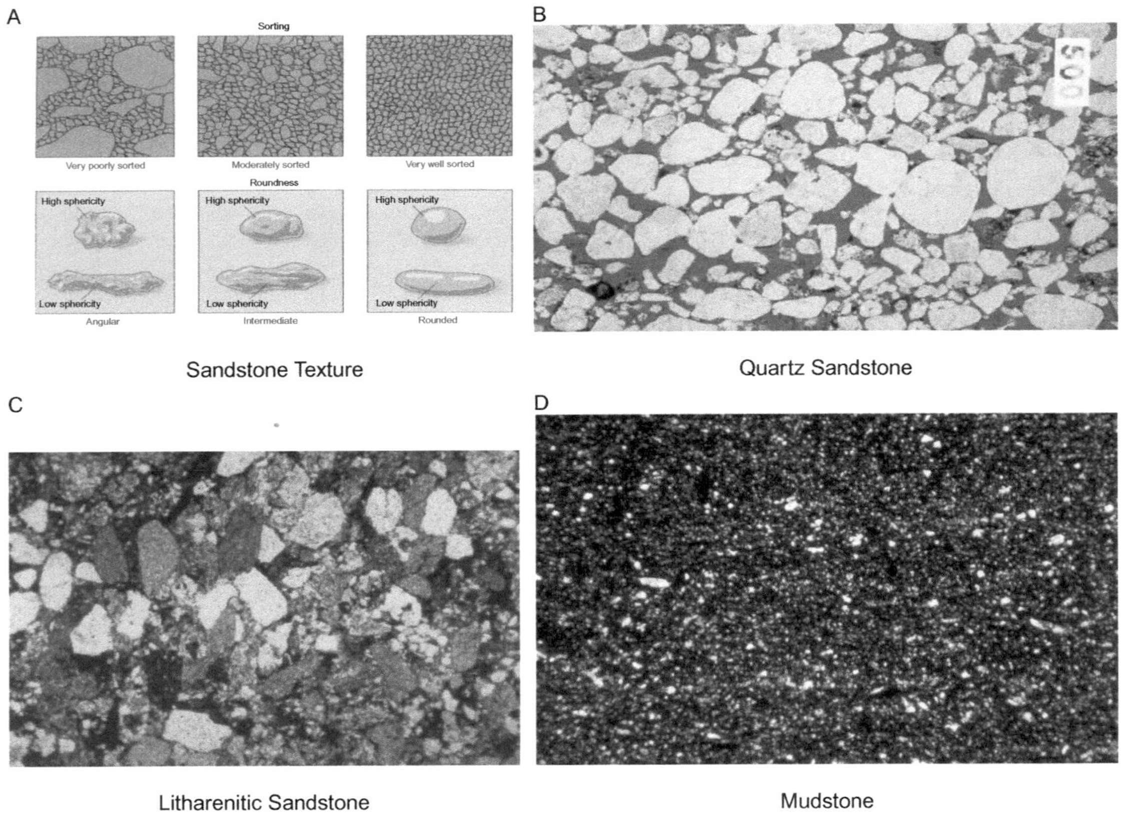

FIGURE 2.5B Clastic Sandstones.

A. Porosity

Porosity is a measure of the volume within a rock that is available to contain reservoir fluids. Therefore, the volume of oil, gas, and water in a given reservoir depends directly on porosity. Porosity (ϕ) is the ratio of the total void space within a rock (the pore volume) to the total bulk volume of that rock. Expressed mathematically:

$$Porosity = \phi = \frac{V_p}{V_B} \tag{2.6}$$

A.1 Types of porosity

1. Primary porosity – formed during initial deposition.
2. Secondary porosity – formed during the stage of diagenesis.
3. Absolute porosity – measure of the total pore spaces in a rock as a function of its bulk volume.
4. Effective porosity – measure of the interconnected pore spaces in a reservoir rock as a function of its bulk volume. This is also a function of the irreducible water saturation. Defined mathematically as:

$$\text{Effective porosity} = \text{porosity}.(1 - S_{wirr}) \tag{2.7}$$

where:

S_{wirr} = Irreducible water saturation

Diagenesis affects the eventual porosity especially in relation to wettability, pack structure, and formation textural properties and pack structure. The structure affects the effective porosity and corresponding pore size distribution (Figure 2.6).

A.2 Porosity Measurement Porosity can be measured directly from representative core samples. The limitations of unconsolidated reservoirs are that good representative core samples are difficult to obtain; selective side-wall core samples whilst being sometimes useful may also lead to poor representation; and reliance on log-derived porosity is now becoming the norm. Porosity can be obtained or derived from sonic, density, neutron, or nucleo-magnetic resonance (NMR) logs.

B. Permeability

This is defined as a measure of the ability of the reservoir rock to transmit a fluid. It is a proportionality constant relating flow rate to pressure change in the rock. It is the most important rock property for production forecasting. Expressed mathematically for a linear flow condition it is given as:

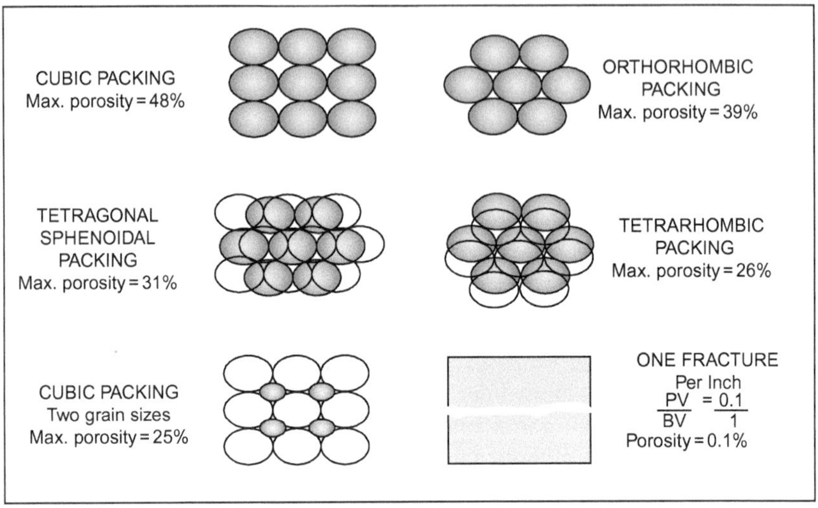

FIGURE 2.6 Effect of Pack Structure on Porosity.

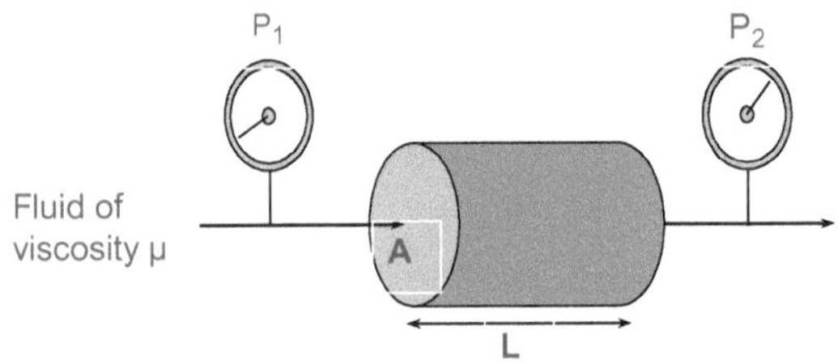

FIGURE 2.7 Typical Permeameter Layout.

$$q = \frac{kA}{\mu}\frac{\Delta p}{\Delta L}$$

$$k = \frac{q\mu\Delta L}{A\Delta p} \tag{2.8}$$

This is called Darcy's law linear flow equation (Fig. 2.7).
where:

q = Flow rate
μ = Fluid viscosity
A = Rock cross-sectional area
Δp = Pressure drop
ΔL = Length

Units are m^2 or Darcy (D) or milliDarcy (mD).

There are different types of permeabilities, discussed as follows.

1. Effective Permeability

When a rock contains more than one fluid the ability of each fluid to flow is impaired by the presence of other fluids. *Effective permeability* is the permeability of a rock to a particular fluid in the presence of a combination of fluids.

where:

k_o = Effective permeability to oil
k_g = Effective permeability to gas
k_w = Effective permeability to water

2. Directional Permeability

There are vertical (k_V), horizontal (k_H), and z-direction (k_Z) permeabilities. Isotropic reservoirs have the same or similar vertical and horizontal permeabilities. Anisotropic reservoirs have different horizontal and vertical permeabilities.

3. Relative Permeability

This is the ratio of effective permeability to a particular fluid as a measure of the absolute rock permeability (k). It is largely dependent on fluid saturation and phase wettability.

where:

$k_{ro} = k_o/k$ = Relative permeability to oil
$k_{rg} = k_g/k$ = Relative permeability to gas
$k_{rw} = k_w/k$ = Relative permeability to water

***Absolute Permeability* Measurement** Absolute permeability can be measured from core samples using a permeameter and computed from Darcy's equation. It is also measured in the reservoir during pressure build-up testing: The value determined is the average effective permeability to oil as seen by testing the presence of initial water saturation. This can be very different than laboratory-derived core permeability.

Application

The absolute permeability is used to determine the relative permeability of different fluids to flow simultaneously in a reservoir.

Permeabilities in Layers

For layered reservoir permeabilities the average permeability can be computed from:

$$k_{avg} = \frac{\sum_{i=1}^{n} k_i h_i}{h_t} \tag{2.9}$$

where:

k = permeability of layer
h_i = thickness of layer
h_t = total thickness

For permeabilities in series (e.g., in a gravelpack completion with a gravel-pack layer, a damaged zone layer, and a lean reservoir permeabilities), the average permeability can be computed with:

$$\Delta p_T = \Delta p_1 + \Delta p_2 + \Delta p_3 = p_1 - p_2$$

$$q_1 = q_2 = q_3 = q$$

$$\therefore k_{avg} = \frac{L_T}{\sum \frac{L_n}{k_n}} \tag{2.10}$$

C. Compressibility

Matrix Compressibility Pore volume compressibility c_f can be defined as the relative change in the pore volume of the rock divided by the change in pressure. It can be expressed mathematically as:

$$c_f = \frac{1}{Vp}\frac{dVp}{dp}\,\text{psi}^{-1} \tag{2.11}$$

where:

V_p = Pore volume
dV_p/dp = Change in pore volume due to change in pressure

Compressibilities for oil, gas, and water can be defined in the same format.

Total Reservoir Compressibility This can be defined as:

$$c_t = c_g s_g + c_o s_o + c_w s_w + c_f \tag{2.12}$$

where:

c_g = Gas compressibility
c_o = Oil compressibility
c_w = Water compressibility
c_f = Matrix or formation compressibility
s_g = Gas saturation
s_o = Oil saturation
s_w = Water saturation

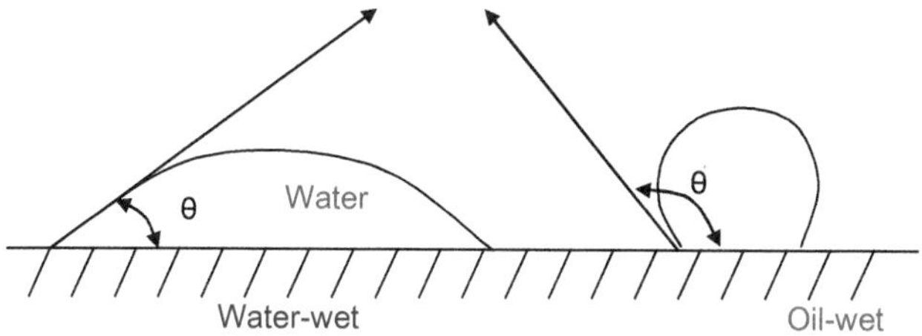

FIGURE 2.8 Wettability of Reservoir Rock.

D. Wettability (Fig. 2.8)

This is a measure of the tendency of one fluid to adhere to the surface of a rock when other fluids are present. Wettability states for reservoirs include water-wet, oil-wet, and intermediate wettability.

E. Capillary Pressure

Capillary pressure is the difference in pressure that exists across any curved interface between two immiscible fluids. In the capillary tube, capillary pressure can be defined as:

$$Pc = \frac{2\sigma_{wo}\text{Cos}\theta}{r}$$

σ_{ow} = interfacial tension between oil and water
θ = contact angle
r = radius of tube

$$Pc = (\rho_w - \rho_o)gh \tag{2.13}$$

$\rho_{w,o}$ = fluid density
g = gravity
h = height of water column

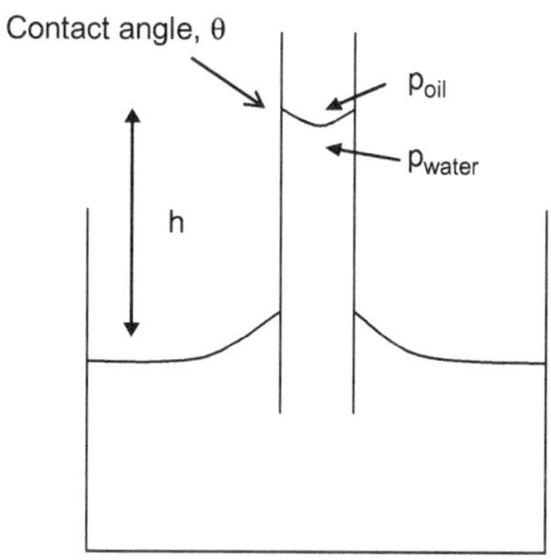

In a *water-wet* oil reservoir there is a transition from 100% water (free-water level) in the water zone to the connate water saturation, S_{wc}, in the oil zone due to capillary pressure. This is known as the transition zone, and its height in field units is:

$$h = \frac{144\,\mathrm{Pc}}{(\rho_w - \rho_o)} \tag{2.14}$$

$Pc = \mathrm{psi}$
$h = \mathrm{ft}$
$\rho_w, \rho_o = \mathrm{lb/cuft}$

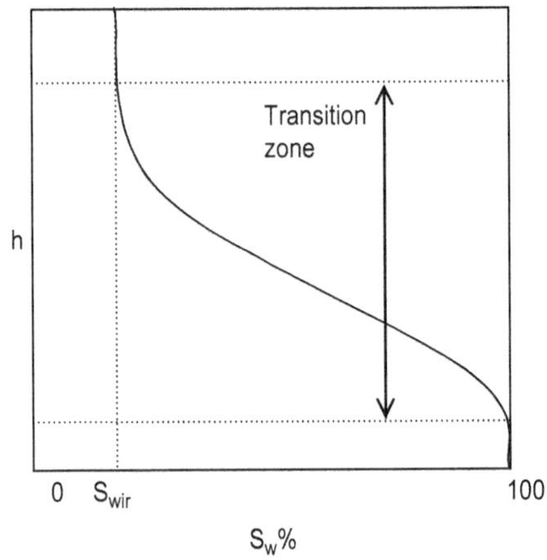

Importance of Capillary Pressure to Sand Production/Management For many unconsolidated reservoirs with grains held together by osmotic pressure, the increase in water encroachment leads to a capillary pressure increase beyond the osmotic pressure leading to a break down of the formation, and corresponding increase, of sand production.

F. Textural Properties

The textural properties are:

- Mean size – This represents the 50th percentile size of the formation grains. Knowledge of formation and sand grain size is required for gravel and screen system selection.
- Size distribution/sorting
- Shape

These properties are important for detailed sedimentological studies and definition of appropriate footprint for sand control design. Traditional percentile size distribution is still utilised for grain size analysis from which sorting and uniformity coefficients are computed using the equations 2.15 to 2.17.

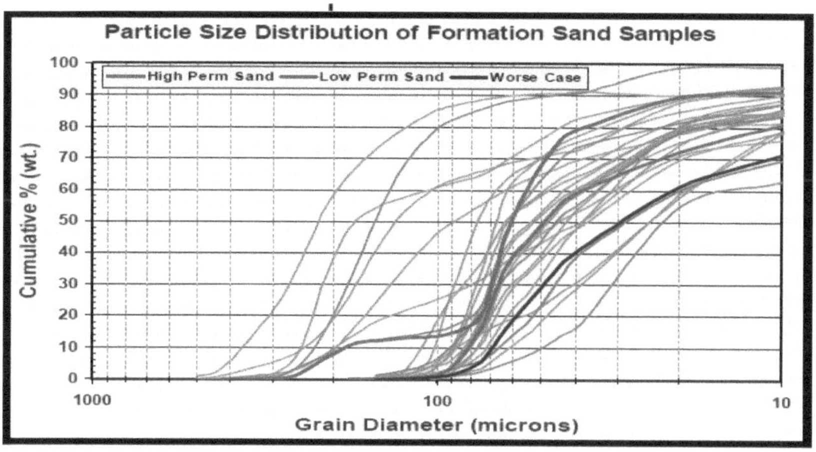

FIGURE 2.9 Formation Sand Grain Size Distribution plot- The Ogive *(Source: http://www.-sandmanagement.com).*

Representative formation samples can be obtained from:

- Logging of cutting samples of pay interval via the standard mud-logging process
- Selective sidewall coring and where possible whole core sampling

Thereafter, detailed analysis can be carried out via:

- Traditional dry sieve analysis
- Use of sophisticated particle analyser: malvern, coulter counter, and quantimet particle analysis machines

A typical grain-size profile is shown in Figure 2.9.

Textural properties can be computed using the following equations:

1. For Sorting

Sedimentological Model:

$$S = \frac{\phi_{84} - \phi_{16}}{4} + \frac{\phi_{95} - \phi_5}{6.6} \tag{2.15}$$

$\phi = -\log_2 d$

$[d] = mm$

Sorting Criteria:

$S < 0.35$ = Very well sorted

$0.35 < S < 0.5$ = Well sorted

$0.5 < S < 0.71$ = Moderately well sorted

$0.71 < S < 1$ = Moderately sorted

$1 < S < 2$ = Poorly sorted

$S > 2$ = Very poorly sorted

Alternatively, the Engineers Method can be used:

$$S = \frac{\phi_{90} - \phi_{10}}{2} \tag{2.16}$$

where:

S < 1 = Well-sorted sand
S > 1 = Poorly sorted

Pore-Size Distribution

This can be computed from the following equations:

$$\begin{aligned} D_{pore} &= \frac{D_{50}}{6.5} \\ &B.Blake - Kozeny \\ D_{pore} &= \frac{D_{50}{}^{*}\phi}{3(1-\phi)} \end{aligned} \tag{2.17}$$

where:

ϕ = Porosity

2.4.3.2 *Elastic or Geomechanical Properties*

The important geomechanical properties determine how much a rock system will deform under a given external or internal pressure. Key properties include:

- Poisson ratio
- Bulk modulus
- Shear modulus
- Compressive strength

Relevance

The properties are very important in:

a. Evaluating wellbore stability/sand failure
b. Sand production rate prediction
c. Perforation strategy

1. Poisson Ratio

This is the ratio of transverse contraction strain to longitudinal external strain in the direction of stretching force. It is positive for tensile deformation and negative for compressive deformation. Mathematically it can be expressed as:

$$\begin{aligned} v &= -\frac{\varepsilon_{trans}}{\varepsilon_{long}} \\ v &= {}^{[3K-2G]}\Big/_{[6K+2G]} \\ E &= 2G[1+v] \end{aligned} \tag{2.18}$$

where:

E = Young's Modulus expressed as:

$$E = \frac{F/A}{\Delta L/L} \tag{2.19}$$

where:

K = Bulk modulus expressed as:

$$K = \frac{\Delta P}{\Delta V/V} \tag{2.20}$$

where:

G = Shear modulus expressed as:

$$G = \frac{FA}{\Delta x/h} \tag{2.21}$$

where:

ε = Represent the transverse and longitudinal strains
F = Force applied per unit cross-sectional area A over a unit length L

2.4.3.3 *Electrical Properties*

These are not direct rock properties but the footprint of the reservoir content. The electrical properties are a measure of voltage resistance and can also be described as the measure of ease with which current will pass through the rock body. The electrical properties are resistivity and conductivity.

Application

- Reservoir fluid identification
- Mud chemistry and formation chemistry

Measurement

- Lab: Resistivity meter
- Field: Electric logs (resistivity log)

Conductivity

- This is for formation water identification and definition of fluid saturation.

Resistivity

- Oil has higher resistivity than formation water.

2.5 RESERVOIR FLUID PROPERTIES

The main reservoir fluids are:

- Heavy oil
- Black oil
- Volatile oil
- High shrinkage oil
- Near critical gas condensate
- Rich gas condensate
- Lean gas condensate
- Wet gas
- Dry gas and
- Formation water

The existence of the fluids in the reservoir or at surface conditions is influenced by the phase behaviour (Figure 2.10) at the different conditions of pressure and temperature. The phase behaviour footprint for each of the hydrocarbon fluids is given in Figure 2.11.

2.5.1 Typical Hydrocarbon Colours

The typical hydrocarbon colours are:

- Heavy oil = Black colour
- Black oil = Brown to dark green
- Volatile oil = Greenish to orange
- Gas condensate = Orange to yellow
- Wet gas or dry gas = Yellow to clear colourless

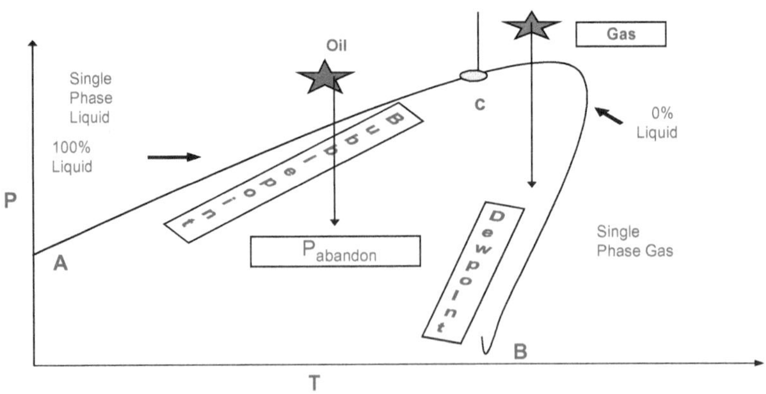

FIGURE 2.10 Phase Envelope for Typical Hydrocarbon Fluids.

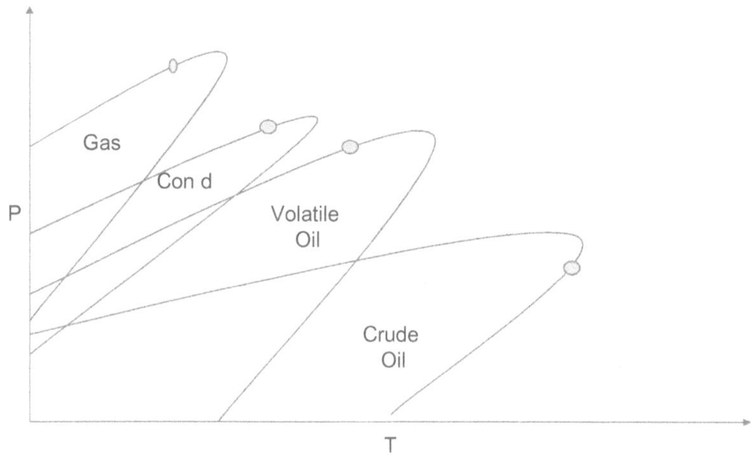

FIGURE 2.11 Phase Envelope for Different Reservoir Fluids.

The key oil and gas properties and formation water properties can be divided into two groups:

1. PVT properties
   a. Formation volume factor
   b. Solution gas-oil ratio, rs
   c. Dissolved gas-oil ratio
   d. Volatilized oil-gas ratio
2. Flow/compositional properties
   a. Density/sg/api gravity
   b. Viscosity
   c. Molecular weight
   d. Compressibility factor
   e. Compressibility
   f. Bubble-point pressure

Oil Formation Volume Factor

- Liquid phase volume at reservoir conditions as a function of the stock tank volume
- Unit = RB/STB
- Symbol = B_o

Gas Formation Volume Factor

- Vapour phase volume at reservoir conditions relative to gas volume at standard conditions
- Units: RB/SCF or MCF/SCF
- Symbol = B_g

Shrinkage Factor = 1/B$_o$ or Expansion Factor = 1/B$_g$
Dissolved Gas – Oil Ratio (R)

- Ratio of the volume of surface gas to stock tank oil in a reservoir liquid phase at reservoir conditions
- Volume of the surface gas and stock tank oil are those measured at standard conditions.
- Unit: SCF/STB

Volatilized Oil-Gas Ratio (R$_v$)

- Volume of stock tank oil to surface gas contained in a reservoir vapour phase at reservoir conditions
- Volume of surface gas and stock tank oil measured at standard conditions
- Unit: STB/MMSCF

These two properties are solubility ratios.

Fluid Density

This is mass per unit cross-sectional area. In practical terms, for oil it can be expressed as specific gravity or API gravity.

$$SG_{oil} = \frac{\rho_{oil}}{\rho_{water}}$$
$$API = \frac{141.5}{SG_{oil}} - 131.5 \tag{2.22}$$

Compressibility Factor

This is the rate of volume occupied by real gas to volume by its ideal equivalent under the same P & T. It varies with changes in composition as well as pressure and temperature. It can also be derived based on the Law of Corresponding States, which specifies that all pure gases have the same compressibility factor at the same values of pseudoreduced pressure and temperature.

A good correlation for z-factor computation is the Hall & Yarborough correlation, given as follows:[4]

$$z = \frac{[1 + x + x^2 - x^3]}{(1 - x)^3} - Ax + Bx^c$$
$$A = 14.76t - 9.76t^2 + 4.58t^3$$
$$B = 90.7t - 242.2t^2 + 42.4t^3$$
$$C = 1.18 + 2.82t \tag{2.23}$$
$$x = \frac{bp}{14zRt}$$
$$b = 0.245\left[RT_c/_{p_c}\right]e^{\left[-1.2(1-t)^2\right]}\ t = \frac{T_c}{T}$$

Other gas properties are:
Gas Density:

$$\rho_{gas} = \frac{pMW}{zR_oT} \tag{2.24}$$

Gas Specific Gravity:

$$\gamma_g = \frac{MW}{28.9966} = \frac{\rho_{gas}}{\rho_{air}} \tag{2.25}$$

Gas Viscosity, μ_g:

$$\begin{aligned}
&\mu_g = 10^{-4} A e^{\left(B \rho_g C\right)} \\
&C = 2.447 - 0.2224B \\
&B = 3.408 + \frac{986.4}{T} + 0.01009MW \\
&A = \frac{(9.379 + 0.01607MW)T^{1.3}}{209.2 + 19.26MW + T} \\
&\left[\mu_g\right] = cp; \ \left[\rho_g\right] = gm/cc \ [T] = {}^{\circ}R
\end{aligned} \tag{2.26}$$

Gas FVF, B_g:

$$B_g = \frac{v_R}{v_s} = \frac{0.005035Tz}{p} \tag{2.27}$$

Mole Fraction of Surface Gas, y_g:

$$y_g = \frac{MW_{osc} - MW_g}{MW_{osc} - MW_{gsc}} \tag{2.28}$$

Volatisiced Gas-Oil Ratio, R_{v}:

$$R_v = \frac{333.3\left[\frac{MW_g}{MW_{gsc}} - 1\right]}{\left[44.29\left(\frac{\gamma_{osc}}{MW_{gsc}}\right) - \left(\frac{MW_g}{MW_{gsc}}\right)(1.03 - \gamma_{osc})\right]} \tag{2.29}$$

Oil SG:

$$\begin{aligned}
&SG_{oil} = \frac{\rho_{oil}}{\rho_{water}} = \gamma_{osc} \\
&API = \frac{141.5}{SG_{oil}} - 131.5
\end{aligned} \tag{2.30}$$

Oil Molecular Weight:

$$For\,API < 40^o \quad MW_{osc} = 630 - 10^*[^oAPI]$$

$$For\,API > 40^o \quad MW_{osc} = 73100^*API^{-1.562} \tag{2.31}$$

$$MW_{osc} = \frac{6084}{^oAPI - 5.9}$$

Solution Gas-Oil Ratio
Example: Standing Correlation

$$R_s = \gamma_g{}^* \left[\frac{p}{18^*10^{y_g}}\right] \tag{2.32}$$

where:

γ_g = Gas gravity, (Air = 1.0)
y_g = Mole fraction gas, 0.00091*T – $0.0125g_{API}$
(p) = psia
(R_s) = SCF/STB
(T) = °F
Oil FVF:

where:

$$B_o = 0.972 + 0.000147^*F^{1.175} \tag{2.33}$$

where:

F = Correlation function, R_s * $(g_g/g_{osc})^*0.5 + 1.25$ T
(T) = Reservoir temperature, °F
γ_g = Gas gravity (Air = 1)

Oil Viscosity
Ng and Egbogah Correlation for Black Oil

$$LOG\,(LOG\,(\mu_{OD} + 1)) = 1.8653 - 0.025086^{*o}API - (0.5644)^*LOG\,(T) \tag{2.34}$$

where:

(T) = Temperature, °F
(μ_{OD}) = Gas-free oil viscosity @ 14.7psia, cp

Beggs and Robinson Correlation
For Gas Saturated Oil

$$\begin{aligned} \mu_o &= A * \mu^B{}_{OD} \\ A &= 10.715(R_s + 100)^{-0.515} \\ B &= 5.44(R_s + 150)^{-0.338} \end{aligned} \tag{2.35}$$

Water Density

$$\rho_W = 62.368 + 0.438603S + 1.60074(E-3)S^2 \tag{2.36}$$

where:

S = Salinity in wt % TDS
TDS = Total dissolved solids
$\rho_{seawater}$ = 64ppcf = 35000 ppm TDS

2.6 FUNDAMENTALS OF RESERVOIR FLUID FLOW

The main equation that defines the flow of fluid through a porous medium (Figure 2.12) is Darcy's Law equation. Consider a porous core sample with a length L and cross-sectional area A through which a fluid of viscosity μ is flowing at a rate q. The flow leads to pressure drop Δp across the core depending on the core permeability, k.

The apparent or superficial velocity v can be mathematically defined as:

$$v = \frac{q}{A} \tag{2.37}$$

$$v = \frac{k}{\mu}\frac{\Delta p}{L} \tag{2.38}$$

where:

k = Permeability, Darcy
μ = Viscosity
ΔP = Pressure drop (P_1-P_2)
L = Length
q = Flow rate
A = Cross-sectional area

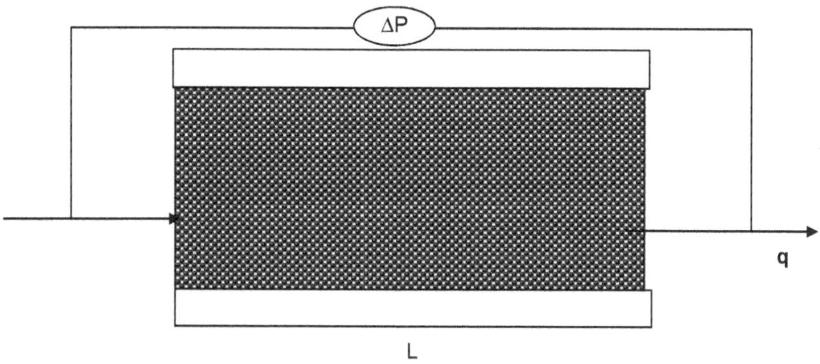

FIGURE 2.12 Schematic of Linear Flow through Porous Medium.

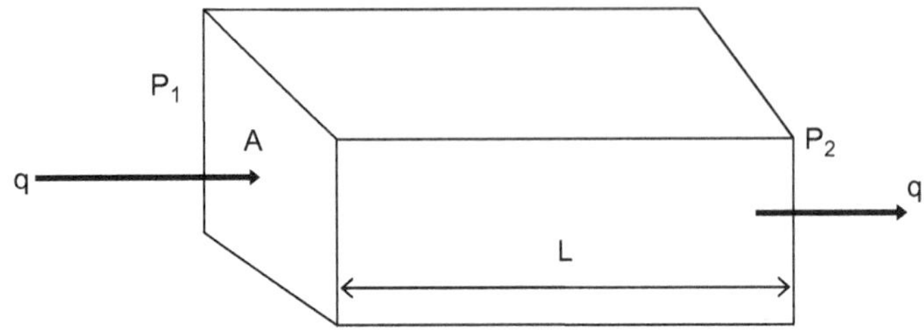

FIGURE 2.13 Linear Flow Model.

2.6.1 Linear Flow Model (Figure 2.13)

Consider a steady-state linear flow of a fluid of a constant viscosity μ through the elemental rectangular core of length L (Figure 2.13). For the linear flow, we have:

$$q = \frac{kA}{\mu} * \frac{(p_1 - p_2)}{L} = \frac{kA}{\mu} * \frac{\Delta p}{L} \tag{2.39}$$

In cgs units:
where:

(q) = Volumetric flow rate, cm^3/s
(A) = Cross-sectional area, cm^2
(L) = Length, cm
(Δp) = Pressure drop, atm
(μ) = Fluid viscosity, cp
(k) = Matrix permeability, Darcy

In field or API units:

$$\mathbf{q} = \frac{1.127kA}{\mu} * \frac{\Delta p}{L} \tag{2.40}$$

q = bbl/day; A = ft^2; L = ft; μ = cp; Δp = psi

$$\mathbf{v} = \frac{7.318k}{\mu} * \frac{\Delta p}{L} \tag{2.41}$$

where:

v = Velocity, ft/sec.
(k) = Darcy

Application Options

This analytical model presented above can be used to analyse core analysis data, especially in relation to evaluating:

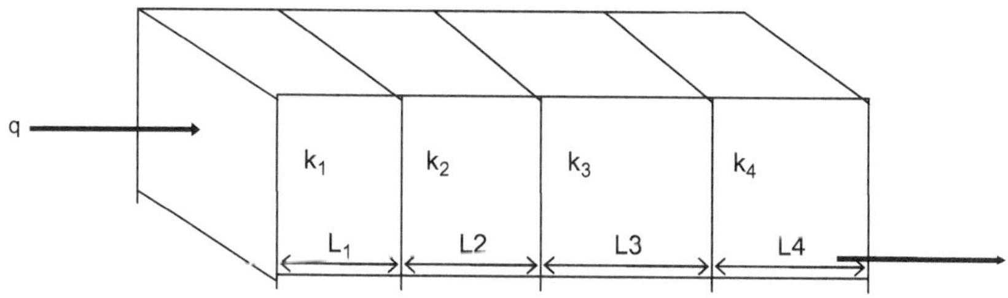

FIGURE 2.14 Linear Flow in Series Model.

(i) Effect of drilling/completion fluid invasion
(ii) Return permeability from flow-back analysis
(iii) Proppant flow capacity analysis

2.6.2 Linear Flow Through Combined Layers

2.6.2.1 Flow in Series (Figure 2.14)

For example, consider a linear steady flow q through four layers of pay sections in series. Assuming the pay interval is uniform with the same cross-sectional area A, then

$$\Delta p_T = \Delta p_1 + \Delta p_2 + \Delta p_3 + \Delta p_4 = p_1 - p_4 = \text{Total Drawdown}$$

$$q = \frac{k_1 A \Delta p_1}{\mu L_1} = \frac{k_2 A \Delta p_2}{\mu L_2} = \frac{k_3 A \Delta p_3}{\mu L_3} = \frac{k_4 A \Delta p_4}{\mu L_4} = \frac{\bar{k} A \Delta p_T}{\mu L_T}$$

$$\Delta p_T = p_1 - p_2 = \frac{q\mu}{A}\left[\frac{L_1}{k_1} + \frac{L_2}{k_2} + \frac{L_3}{k_3} + \frac{L_4}{k_4}\right] = \frac{q\mu}{A} * \frac{L}{\bar{k}} \quad (2.42)$$

$$\therefore \frac{L}{\bar{k}} = \left[\frac{L_1}{k_1} + \frac{L_2}{k_2} + \frac{L_3}{k_3} + \frac{L_4}{k_4}\right]$$

Application Options

This model can be used to analyse:

(i) Effect of mechanical damage due to invasion on the permeability of a reservoir rock for typical core flood drilling fluid screening risk assessment study
(ii) Depth of invasion of drilling and completion fluid
(iii) Effect of gravelpack on reservoir inflow
(iv) Contributions by each reservoir section for production logging data analysis

2.6.2.2 Flow in Parallel (Figure. 2.15)

In this case, the pressure drop through each parallel layer is the same. However, the flow rate through each layer is a function of the individual layer permeability.

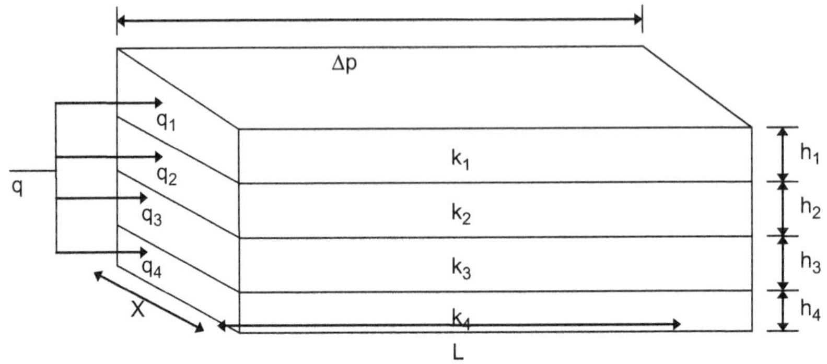

FIGURE 2.15 Multilayer Flow Model.

$$q = q_1 + q_2 + q_3 + q_4$$

$$q = \frac{k_T h_T X}{\mu} \frac{\Delta p}{L} = \frac{k_1 h_1 X}{\mu} \frac{\Delta p}{L} = \frac{k_2 h_2 X}{\mu} \frac{\Delta p}{L} = \frac{k_3 h_3 X}{\mu} \frac{\Delta p}{L} = \frac{k_4 h_4 X}{\mu} \frac{\Delta p}{L} \tag{2.43}$$

$$k_T h_T = k_1 h_1 + k_2 h_2 + k_3 h_3 + k_4 h_4$$

2.6.3 Radial Flow Model (Figure 2.16)

This case represents the most practical condition of flow into the wellbore, especially in open-hole completions.

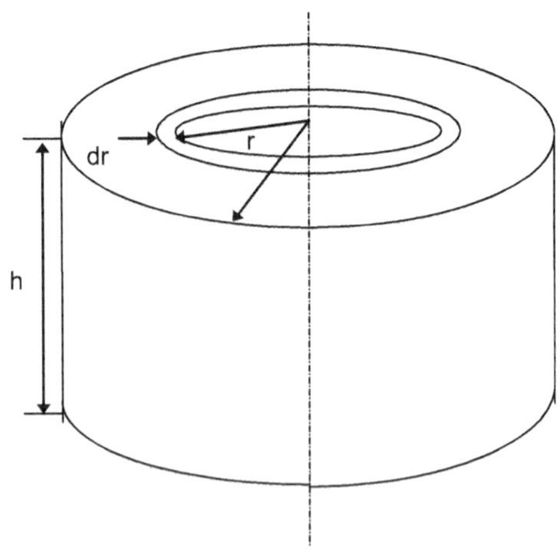

FIGURE 2.16 Radial Flow Model.

Consider radial flow in a horizontal direction through an elemental system dr at a radius r from the centre. Assuming steady-state flow, we have from the Darcy's law equation:

$$q=\frac{kA}{\mu}*\frac{dp}{dr} \qquad A=2\pi rh$$

$$\frac{q}{2\pi kh}\int_{r_1}^{r_2}\frac{dr}{r}=\int_{p_1}^{p_2}dp$$

$$\therefore q=\frac{2\pi kh}{\mu}*\frac{\Delta p}{\ln\frac{r_2}{r_1}} \tag{2.44}$$

In field or API units:

$$q=\frac{7.08kh}{\mu}*\frac{\Delta p}{\ln\frac{r_2}{r_1}} \tag{2.45}$$

where:

k = Darcy
μ = cp
Δp = psi
h = ft
q = bbls/day
r = ft
r_2 = Outer radius and r_1 represents the inner radius
q = Reservoir delivery to wellbore, bbl/day
$q_{s,t}$ = q/B, stock tank production rate
(q_s) = Stock tank barrels per day, STB/day
B = Formation volume factor, rb/stb

For an oil-producing well, for example:

$$q_{so}=\frac{7.08k_oh}{\mu_o}*\frac{[P_R-P_{wf}]}{\ln\frac{r_e}{r_w}} \tag{2.46}$$

where:

(q_{so}) = STB/day of oil production
k_o = Reservoir effective permeability to oil, Darcy
μ_o = Oil viscosity, cp
h = Pay thickness, ft
P_R = Reservoir pressure, psi
P_{wf} = Flowing bottom-hole pressure, psi
r_e = Reservoir drainage radius, ft
r_w = Wellbore radius

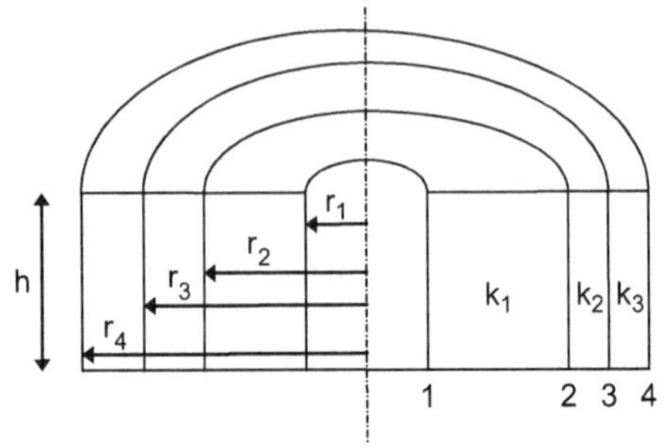

FIGURE 2.17 Radial Flow in Series.

2.6.3.1 Radial Flow in Series (Figure 2.17)

As in the case of linear flow, it can be proved that:

$$q = \frac{2\pi k_2 h(p_2 - p_1)}{\mu \ln\left(\frac{r_2}{r_1}\right)} = \frac{2\pi k_2 h(p_3 - p_2)}{\mu \ln\left(\frac{r_3}{r_2}\right)} = \frac{2\pi k_3 h(p_4 - p_3)}{\mu \ln\left(\frac{r_4}{r_3}\right)} \tag{2.47}$$

$$q = \frac{2\pi \bar{k} h}{\mu} * \frac{\Delta p}{\ln \frac{r_4}{r_1}}$$

$$\therefore \frac{\ln\left(\frac{r_4}{r_1}\right)}{\bar{k}} = \frac{\ln\left(\frac{r_2}{r_1}\right)}{k_1} + \frac{\ln\left(\frac{r_3}{r_2}\right)}{k_2} + \frac{\ln\left(\frac{r_4}{r_3}\right)}{k_3}$$

where p_1, p_2, p_3, and p_4 are pressures at boundaries 1, 2, 3, and 4 respectively.

Application Options

This model can be used to analyse:

(i) Effect of mechanical damage due to invasion on the permeability of a reservoir rock.
(ii) Depth of invasion of drilling and completion fluid
(iii) Effect of permeability profile on reservoir inflow

2.6.3.2 Radial Flow in Parallel (Figure 2.18)

For radial flow in parallel, it can be proven that:

$$\bar{k} h_T = k_1 h_1 + k_2 h_2 + k_3 h_3 \tag{2.48}$$

The above equations are valid for steady-state incompressible fluid flow. For compressible flow, the mass flow rate can be said to be constant but the volumetric flow rate changes with the conditions of temperature and pressure as well as fluid composition.

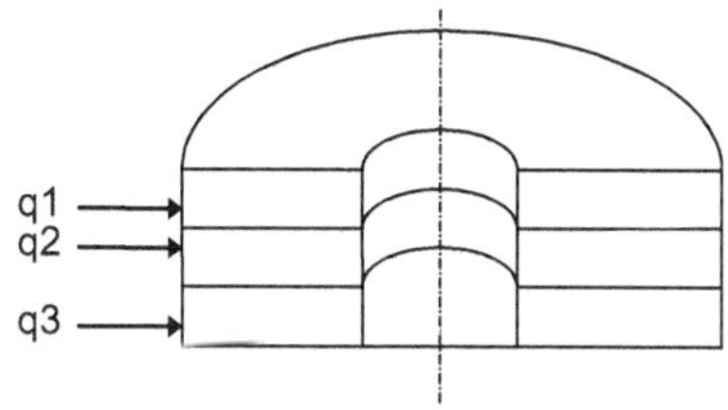

FIGURE 2.18 Radial Flow in Parallel.

Application Options

This model can be used to analyse:

(i) Effect of delivery from each section of a multilayer reservoir as part of production logging operation
(ii) Effect of heterogeneity profile on reservoir performance

2.6.4 Compressible Fluid Flow

Gas flow is a typical example of flow through the porous medium. For such flow, the flow rate q is the mean flow rate expressed as:

$$q_m = \frac{2\pi k_g h \Delta p}{\mu \ln \frac{r_2}{r_1}} \tag{2.49}$$

where:

q_m = Mean volumetric flow rate at mean flowing pressure, p_m

$$p_m = (p_1 + p_2)/2 \tag{2.50}$$

From the gas law relationship, the flow rate at standard conditions can be computed as:

$$\frac{q_{sc} p_{sc}}{z_{sc} T_{sc}} = \frac{q_m p_m}{z_m T_m}$$
$$\therefore q_{sc} = \frac{2\pi k_g h z_{sc} T_{sc} p_m (p_1 - p_2)}{p_{sc} z_m T_m \mu_g \ln \frac{r_2}{r_1}}; \tag{2.51}$$

In API Units:

$$q_{sc} = \frac{703 k_g h \left(p_R^2 - p_{wf}^2\right)}{T_m z_m \mu_g \ln \frac{r_e}{r_w}} \tag{2.52}$$

where:

(q_{sc}) = SCF/day
k_g = Effective permeability to gas, Darcy
h = Pay thickness ft
(p) = psi
(r) = ft

2.6.5 Drainage and Imbibition Processes

For a drainage process, the wetting phase saturation is continuously decreasing (as in a gas drive for a water wet reservoir), while in the imbibition process the wetting phase is continuously increasing (as in water encroachment). Imbibition is a source of formation damage as it may lead to reduction in relative permeability to hydrocarbons.

2.7 PRODUCTIVITY AWARENESS

2.7.1 Reservoir Production Concepts

A reservoir rock will produce naturally as a consequence of the fluid it contains, existing at high pressure, and the rock under continuous state of compaction. The reservoir contains a high level of energy in the form of compression force, which can be utilised for fluid production from the reservoir into the wellbore and to the surface all the way to the treatment facilities.

The response of the reservoir to the depletion process is dynamic, since the reservoir fluid is continuously changing in terms of volume, composition, and other properties. The response of the reservoir system to depletion is governed by the reservoir production mechanism.

2.7.2 Reservoir Depletion Response

The basic concept regarding the production of fluid from a reservoir is that for fluid to be produced as a result of its high pressure there must be pressure drop. To maintain the pressure, the reservoir must compensate for the loss of the produced fluid by one of the following mechanisms:

1. Expansion of the reservoir rock matrix
2. Expansion of the connate water
3. Expansion of the hydrocarbon phases present in the reservoir including:
   - Expansion of undersaturated oil if pressure is above bubble-point
   - Expansion of oil/gas if pressure is below bubble-point
   - Expansion of the overlying gas cap
4. Expansion of the underlying aquifer

In most cases, as the reservoir is depleted, the system cannot maintain its pressure and the overall pressure in the reservoir will decline. The utilisation of the reservoir pressure was discussed in the section on the composite production system. Effective utilisation of reservoir energy for improved production can be attained either by:

(i) increasing reservoir pressure through fluid injection into the reservoir (which can include water or gas injection); or
(ii) providing more energy for the vertical lift process. Examples include gas lift, down-hole pumping using electric submersible pumping (ESP), hydraulic down-hole pumping, sucker rob pumping, or jet pumping.

2.7.3 Reservoir Flow Dynamics

There are three basic types of flow describing the flow dynamics in a reservoir:

1. Unsteady-state flow
2. Steady-state flow
3. Semi-steady state flow

1. *Unsteady-state Flow*

As a reservoir is depleted, the volume is continuously changing. This is accompanied by continuous change in fluid composition, properties, pressure, and interaction with the host rock. This phenomenon represents an unsteady-state situation, which is called transient phenomena, as discussed. Although this is an ideal situation, in normal practice, over specified periods, fluids can be produced at specified rates from a defined drainage radius. This applies during the relatively short period taken by the effect of pressure disturbance occurring at the wellbore to reach the boundary of the reservoir. During this period, the reservoir responds to the disturbance as if it is of infinite extent.

2. *Steady-state Flow*

As fluid flows across the drainage radius, if the volumetric flow rate into the reservoir is equal to the production rate of fluids from the reservoir, the reservoir can be said to be in steady-state flow. This is a condition whereby pressure and flow conditions at every point in the reservoir do not change with time. There is mass flow into the reservoir to maintain pressure as fluid is produced. Example cases include: Large gas cap, strong aquifer, water injection, and gas reinjection.

3. *Semi-Steady State Flow*

If no fluid flows across the outer boundary of the reservoir during production, the fluid production can be compensated for by the expansion of residual fluids in the reservoir. In such a situation, production will cause a reduction in pressure throughout the reservoir unit. This represents a pseudo-steady state or semi-steady state situation. This applies to a reservoir that has been producing at a constant rate for a sufficiently long period for the effect of the outer boundary to become apparent. Conditions are not truly steady state since pressure decreases with time at every point in the reservoir but the rate of change of pressure is independent of time. Schematically, the different stages in the life of a reservoir can be illustrated as shown in Figure 2.19.

Based on the above, there exists a critical time below which we are in transient zone (i.e., unsteady-state flow condition). Above this critical time, the reservoir will be in the 'steady' region, which can be a steady-state flow condition or a semi-steady state (otherwise known as a pseudo-steady state).

The critical time (days) is defined as:

$$t_c = \frac{r_e^2}{4\eta} \tag{2.53}$$

where:

r_e = Reservoir drainage radius, ft
η = Diffusivity constant defined as:

$$\eta = \frac{6.33k}{\phi\mu c} \tag{2.54}$$

where:

k = Formation permeability, Darcy
ϕ = Formation porosity
μ = Fluid viscosity, cp
c = Compressibility, psi^{-1}

For Unsteady-state Flow (t < tc)

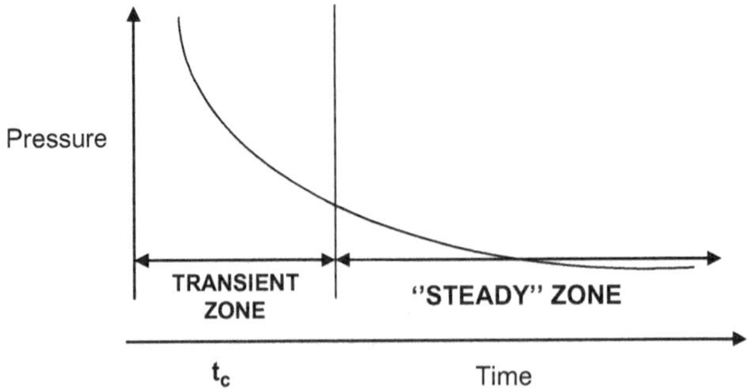

FIGURE 2.19 Pressure Distribution Profile vs. Time.

The pressure distribution equation is given as:

$$p = p_i + \frac{q\mu}{14.16kh} Ei\left[-\frac{r^2}{4\eta t}\right] \tag{2.55}$$

where:

E_i = Epsilon integral
p = Pressure in the reservoir at radius r from wellbore centre
q = Reservoir production rate, bpd
k = Reservoir permeability, Darcy

Application Option

This is very useful for well test data interpretation.

'Steady' Flow Region

If t > tc, the reservoir fluid is in the 'steady' region. In this region, flow can be in the steady state or semi-steady state depending on the operating conditions defined earlier.

For Steady-state Flow

For Liquid Flow

$$q = \frac{7.08kh}{\mu} \frac{p_R - p_{wf}}{\ln \frac{r_e}{r_w} + s} \tag{2.56}$$

where:

q = Reservoir production rate, bpd
k = Effective liquid permeability, Darcy
h = Pay thickness, ft
p_R = Reservoir pressure, psi
p_{wf} = Flowing bottom-hole pressure, psi
r_e = Reservoir drainage radius
r_w = Wellbore radius
s = Total skin effect = measure of damage

For Gas Reservoir in Steady State

$$q_{sc} = \frac{703k_g h\left(p_R^2 - p_{wf}^2\right)}{T_m z_m \mu_g \left[\ln \frac{r_e}{r_w} + s\right]} \tag{2.57}$$

where:

q_{sc} = SCF/day of production
k_g = Effective gas permeability, Darcy

T_m = Average reservoir temperature, °R
z_m = Average gas compressibility factor at $p_m = \frac{p_R + p_{wf}}{2}$ and T_m
μ_g = Average gas viscosity
s = Total skin = $s_d + s_p + spp + s_\theta + s_s + s_g$ +, etc.
s_d = Invasion skin or mechanical skin given as $s_d = \frac{k - k_d}{k_d} \ln \frac{r_d}{r_w}$
s_p = Perforation skin
s_{pp} = Partial penetration skin
s_c = Perforation compaction skin
s_s = Skin due to change in effective stress
s_g = Gravelpack skin
s_θ = Skin due to effect of hole angle

For Semi-steady State Flow Condition

For Liquid

$$q = \frac{7.08kh}{\mu} \frac{p_R - p_{wf}}{\ln \frac{r_e}{r_w} - 0.5 + s} \tag{2.58}$$

For Gas Reservoir in Semi-steady State

$$q_{sc} = \frac{703k_g h \left(p_R^2 - p_{wf}^2\right)}{T_m z_m \mu_g \left[\ln \frac{r_e}{r_w} - 0.5 + s\right]} \tag{2.59}$$

2.7.4 Volume-Averaged Pressure

When a reservoir is drained by a number of wells all operating under semi-steady state conditions, each well will drain within its no-flow boundary. The rate of pressure decline will then be the same at every point within the reservoir. Therefore, the average reservoir pressure can be expressed as:

$$\bar{p}_{res} = \frac{\sum_i \bar{p}_i V_i}{\sum_i V_i} = \frac{\int_{r_w}^{r_e} p dV}{\int_{r_w}^{r_e} dV} \qquad dV = 2\pi r h \phi dr \tag{2.60}$$

Therefore, for example, for semi-steady state under this condition:

$$q = \frac{7.08kh}{\mu} \cdot \frac{(\bar{p} - p_{wf})}{\left[\ln \frac{r_e}{r_w} - \frac{3}{4} + s\right]} \tag{2.61}$$

For steady state under this condition of average pressure, the flow equation is given as:

$$q = \frac{7.08kh}{\mu} \cdot \frac{(\bar{p} - p_{wf})}{\left[\ln\frac{r_e}{r_w} - \frac{1}{2} + s\right]} \tag{2.62}$$

where:

$\bar{p}$ = *average reservoir pressure*

2.7.5 Evaluation of Well Performance

In general, the Darcy's Law equation as applied to laminar flow is given as:

For Radial Flow

For Steady State Flow

$$q_s = \frac{7.08\,kh}{B\ \ \mu} \cdot \frac{\Delta p}{\ln\frac{r_e}{r_w}} \tag{2.63}$$

For Semi-steady State Incompressible Flow

$$\begin{aligned} q_s &= \frac{7.08\,kh}{B\ \ \mu} \cdot \frac{\Delta p}{\ln\frac{r_e}{r_w} - \frac{1}{2}} \quad for \quad \Delta p = p_R - p_{wf} \\ q_s &= \frac{7.08\,kh}{B\ \ \mu} \cdot \frac{\Delta p}{\ln\frac{r_e}{r_w} - \frac{3}{4}} \quad for \quad \Delta p = \bar{p} - p_{wf} \end{aligned} \tag{2.64}$$

where:

q_s = Stock tank production rate, stb/day

Note that the formation volume factor

$$B = \frac{q}{q_s}$$

For such systems, the production performance is measured in terms of three distinct aspects:

1. Well Inflow performance – Flow from reservoir to wellbore
2. Vertical lift performance – Flow up the production string to surface
3. Choke or bean performance – Flow through control systems to reduce flowing pressure

2.7.5.1 Well Inflow Performance

The well production or injection rate is related to the flowing bottom-hole pressure by the so-called inflow performance relationship (IPR). For single-phase flow of oil or gas or water, the relationship is as originally expressed in section 2.6.4. However, for a two-phase flow, there are several correlations but a typical example is the Vogel correlation given as:

$$\frac{q_o}{q_{o\max}} = 1 - 0.2\frac{p_{wf}}{p_R} - 0.8\left[\frac{p_{wf}}{p_R}\right]^2 \tag{2.65}$$

The Vogel correlation is very useful to predict the oil production rate when free gas flows, although only the oil production rate is employed.

$$q_{o\max} = \frac{AOFP}{1.8} \tag{2.66}$$

where:

AOFP = Absolute open flow potential for single phase oil flow

For an undersaturated reservoir with $P_R > Pb$ the Vogel correlation becomes:

$$q_o = q_b + (q_{o\max} - q_b)\left[1 - 0.2\frac{p_{wf}}{p_b} - 0.8\left[\frac{p_{wf}}{p_b}\right]^2\right] \tag{2.67}$$

Another popular correlation is the Fetkovitch correlation, which can be expressed as:

$$q_o = C\left[\bar{p}_R^2 - p_{wf}^2\right]^n \tag{2.68}$$

where:

C = Flow coefficient
n = Exponent depending on well characteristics

2.7.5.2 *Productivity Index*

Well inflow performance is expressed in terms of the productivity index. The productivity index (PI or J) is a measure of the capability of a reservoir to deliver fluids to the bottom of a wellbore for production. It defines the relationship between the surface production rate and the pressure drop (drawdown) across the reservoir. Expressed mathematically, it is given as:

$$\mathrm{PI} = \mathrm{J} = \frac{q_s}{P_R - p_{wf}} \tag{2.69}$$

For Steady-state Flow of Incompressible Fluid

$$\mathrm{PI} = \frac{q_s}{p_e - p_{wf}} = \frac{7.08kh}{\mu B\left[\ln\frac{r_e}{r_w} + s\right]} \tag{2.70}$$

For Semi-steady State

$$PI = \frac{q_s}{\bar{p} - p_{wf}} = \frac{7.08kh}{\mu B\left[\ln\frac{r_e}{r_w} - \frac{3}{4} + s\right]} \quad for \ average \ pressure$$

$$PI = \frac{q_s}{p_e - p_{wf}} = \frac{7.08kh}{\mu B\left[\ln\frac{r_e}{r_w} - \frac{1}{2} + s\right]} \quad for \ normal \ pressure \tag{2.71}$$

For Gas Well

$$PI = \frac{q_s}{p_e^2 - p_{wf}^2} = \frac{703kh}{T(\mu z)_{ave}\left[\ln\frac{r_e}{r_w} + s\right]}{}' scf/psi^2 \tag{2.72}$$

(k) = Permeability, Darcy

Effect of Damage

The above equations have been based on the ideal situation of no damage to the reservoir due to completion, etc. Formation damage can be said to be a measure of the reduction in the production capacity of a reservoir. It is expressed in terms of skin factors.

Sources of damage include:

1. Damage around the near wellbore due to drilling/completion fluid invasion, s_d
2. Damage due to perforations, s_p
3. Damage due to partial completions, s_{pp}
4. Damage due to perforation compaction, s_c
5. Damage due to heterogeneity
6. Damage due to hole angle, etc.
7. Damage due to stress, s_s

The total skin = sum of all skins.

$$s_T = s_d + s_p + s_{pp} + s_c + \cdots \tag{2.73}$$

Therefore, these conditions create a change in the net flow bottom-hole pressure (Fig. 2.20):

$$P_{wfactual} < P_{wfideal}$$

$$P_{wfideal} = P_{wf\,actual} + \Delta P_{skin}$$

$$s = \frac{7.08kh}{q\mu}\Delta p_{skin} \tag{2.74}$$

For near wellbore invasion, $s_d = \frac{k - k_d}{k_d}\ln\frac{r_d}{r_w}$.

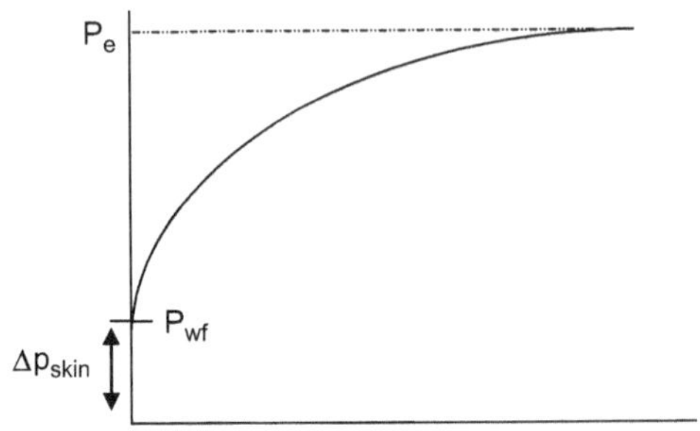

FIGURE 2.20 Pressure Profile in the Reservoir.

where:

k_d = Damaged zone permeability
k = Original formation permeability
r_d = Invaded zone radius
r_w = Wellbore radius

Flow Efficiency $= \dfrac{J_{actual}}{J_{ideal}}$ $J_{ideal} = J$ at $s = 0$

Flow efficiency is normally less than 1.

Damage Factor = 1/FE

For stimulation, we consider productivity ratio, PR, or FOI (folds of increase): PR or FOI $= \dfrac{J_{after}}{J_{before}}$ $J_{after} = J$ after stimulation; $J_{before} = J$ before stimulation. FOI is usually greater than 1.

2.8 PRODUCTIVITY IN HORIZONTAL WELLS

A number of correlations are available for the analysis of horizontal well performance depending on whether we are dealing with an isotropic ($k_v = k_H$) or anisotropic reservoir. Typical examples of horizontal well correlations are Borisov, Babu and Odeh, Giger, Joshi, Economidies, etc.

A typical example of such correlations by Joshi/Economidies is given as:

$$q_s = \frac{7.08 k_H h \left[p_R - p_{wf}\right]}{B\mu \left[\ln \left\{ \dfrac{\left[a + \sqrt{a^2 - \left(\dfrac{L}{2}\right)^2} \right]}{\dfrac{L}{2}} \right\} + \dfrac{I_{ani} h}{L} \ln \dfrac{I_{ani} h}{\left[r_w (I_{ani} + 1) \right]} \right]} \tag{2.75}$$

$$I_{ani} = \sqrt{\frac{k_H}{k_V}}$$

$$a = \frac{L}{2}\left[0.5 + \sqrt{\left(0.25 + \left(\frac{r_{eH}}{L/2}\right)^4\right)}\right]^{0.5}$$

where:

r_{eH} = Equivalent radius in a presumed circular shape of a given drainage area

An example of the PI equation for a damaged horizontal well is given by Renard & Dupuy[7] as:

$$PI_h = \frac{7.08 k_H h / (\mu_o B_o)}{\cosh^{-1} X + \left(\frac{\beta h}{L}\right) \ln\left[h/2\pi r'_w\right] + s_h} \quad (2.76)$$

where:

$X = 2a/L$
$r_w{}'$ = Effective wellbore radius (a number of correlations are available to compute this)

$$s_h = \left(\frac{\beta h}{L}\right)\left[\frac{k}{k_d} - 1\right] \ln\frac{r_d}{r_w} = \left(\frac{\beta h}{L}\right) s_v \quad (2.77)$$

where:

s_v = Vertical well damage

2.9 NODAL ANALYSIS

Nodal analysis is a very powerful tool used in the oil industry to synchronise inflow from the reservoir with tubing production in the wellbore. Basically, a look at the composite production system in Figure 2.1 shows that for a given wellhead pressure, P_{TH}, tubing performance allows the calculation of flowing bottom-hole pressure to lift a range of production rates. Combining the tubing performance with an inflow performance from the reservoir rock (IPR) provides the well deliverability at the determined flowing bottom-hole pressure. This can be done for different tubing sizes.

The steps involved in this analysis are:

2.9.1 Establishment of IPR Curve (Figure 2.21)

The recommended equation for this case, for simplicity, is the Vogel correlation, given as:

$$\frac{q_o}{q_o\max} = 1 - 0.2\frac{p_{wf}}{p_R} - 0.8\left[\frac{p_{wf}}{p_R}\right]^2 \quad (2.78)$$

$$q_o\max = \frac{AOFP}{1.8} \quad (2.79)$$

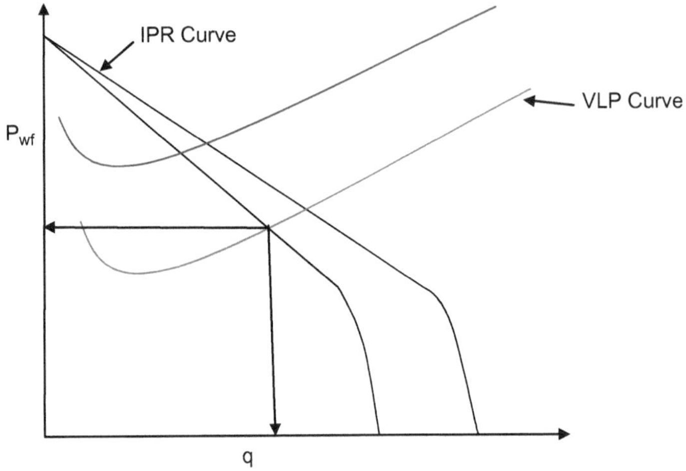

FIGURE 2.21 Nodal Analysis Curves.

where:

AOFP = Absolute open flow potential for single phase oil flow, q_{max} at $p_{wf} = 0$ psig or 14.7 psia

For an undersaturated reservoir with $P_R > P_b$ the Vogel correlation becomes:

$$q_o = q_b + (q_{omax} - q_b)\left[1 - 0.2\frac{p_{wf}}{p_b} - 0.8\left[\frac{p_{wf}}{p_b}\right]^2\right] \quad (2.80)$$

where:

q_o = Production rate, stb/day
p_{wf} = Flowing bottom-hole pressure, psi
p_b = Bubble-point pressure, psi

2.9.2 Tubing Performance Correlation

$$P_{wf} - P_{TH} = \Delta P_{HYD} + \Delta P_{KE} + \Delta P_{FRIC} \quad (2.81)$$

where:

$\Delta P_{HYD} = \rho * g * TVL_H + \rho * g * TVL_T$
TVL_H = True vertical length of hector
TVL_T = True vertical length of tubing minus hector
ρ = Density of fluid

$$\Delta P_{KE} = \frac{\rho v_H^2}{2}\left[\left(\frac{A_T}{A_H}\right)^2 - 1\right] \tag{2.82}$$

2.9.3 Frictional Pressure Drop

The following equations have been adopted for friction loss analysis:

$$\Delta p = \frac{f_{MR} L \rho v^2}{25.8d} \tag{2.83}$$

where:

(Δp) = psi
(f_{MR}) = Friction factor
(L) = Tubing length, ft
(v) = Average fluid flow velocity in tubing, ft/sec
(d) = Tubing ID, ins
(ρ) = Produced fluid density, ppg

2.10 FLOW ASSURANCE MANAGEMENT

The petroleum industry is sourcing advanced technologies to enhance production and recovery from existing mature and marginal fields with more than 50% of recoverable reserves still in place, and carrying out aggressive exploration to new frontiers such as ultra deepwater environments and arctic regions. The deepwater environment (from 2000 m to over 3000 meters), which includes major assets in the the Gulf of Guinea in the West African subregion, the Campos Basin in South America, and the arctic regions, now form the cornerstone of many future oil/gas field developments. Deepwater environments are characterised by very hostile environmental sea conditions and very complex temperature and geological features such as high-pressure high-temperature (HP-HT) with a very narrow window of rock pore pressure-fracture pressure, which makes drilling by conventional overbalanced methods difficult and creates shallow unconsolidated reservoir sands with massive sand production problems and a combination of challenging and 'flexible' rock lithologies.

The economic viability of the mature and marginal fields, deepwater and ultra deepwater environments, and even the arctic regions requires the depletion of these reservoirs with a minimum number of wells. To achieve this, there is an increasing trend toward the use of long horizontal, extended reach, and multilateral wells and long sub-sea tiebacks. In addition to trying to minimise problems associated with fluid production there is the added benefit of using these wells to potentially reduce field development costs.

The major challenges for companies are in how to maximise individual wells and field development performance, minimise operational cost and non-productive time, and guarantee flow assurance across the composite production system from reservoir through the wellbore and from the long sub-sea tieback to the topside for effective asset integrity.

The new frontiers are getting deeper, with many companies now operating at more than 3000 m of water depth, especially in deepwater Gulf of Mexico, Gulf of Guinea, and Campos Basin, far from existing infrastructures. In these environments the reservoirs are more complex and dominated by unconsolidated sands at very high pressure and high temperature. At 2000 m of water depth and above, for economic viability sub-sea production systems become inevitable with sub-sea processing becoming popular in the development along with the use of long distance satellite tiebacks to existing infrastructures. Lengths of 250 km and above of sattelite tiebacks now exist or are being planned for many of the major deepwater frontiers. The sub-sea systems operate as multiphase systems with inherent potential for phase segregation and slugging. Engineers also need to drill and produce more complex wells through challenging high-pressure-high temperature lithologies at the edge of the operating window and will need to push the limit of tolerance in these deepwater environments. The challenges are in how to maximise field development performance and guarantee flow assurance.

The deepwater environment can be categorised into three main depth tiers; namely:

- Deep – Tier I
  - 500-2000 m of water depth
- Very Deepwater Depths – Tier II
  - 2000-3000 m of water depth. These represent the current new exploration frontiers.
- Ultra Deepwater – Tier III
  - Greater than 3000 m of water depth

At these water depths, the sea-bed temperatures can be as low as 100 °F, with major challenges for flow assurance with cold flow of produced fluid and impacts on material integrity. In these environments the lithologies are more complex, with variable pore pressure gradients and very narrow pore pressure-frac pressure window. They are dominated by very high-pressure high-temperature unconsolidated reservoirs with the potential for rapid depressurisation, sand production, and very early water breakthrough and major multiphase fluid production issues. Many of the reservoirs boast of pressure greater than 20000 psi and temperatures greater than 350 °F.

Typical flow assurance issues associated with production covered here include:

- Scales and hydrates formation/management
- Alsphatene and paraffin wax formation/management
- Emulsion blocking
- Slugging and phase separation
- Erosion and corrosion management
- Of course, sand production/management – the main focus of this book
- Produced sand transport in multiphase environment

The many mature well/field and especially sub-sea systems operate as a multiphase systems (gas, water, oil, and solids) with inherent potential for solids deposition, hydrate, and scale formation, severe slugging, transient multiphase flow pattern and flow regime changes complicated by inherent startup and blowdown conditions with corresponding pressure and temperature-profile changes along the tieback, and heat transfer/cooling.

For example, production through a well-in-production completion tubing and the loading of the tiebacks from the individual wells and flow of the produced fluids along the tieback to the beach are accompanied by pressure drop and expansion of the compressible fluids with temperature dropping along the tieback aggravated by rapid 3-D heat transfer across the pipeline from the hot produced fluid and the extremely cold sea-bed conditions. Adsorbed water within the gas tends to condense, leading to the formation of complex multiphase fluid components in the pipeline; namely:

1. Formation water
2. Condensed water
3. Gas
4. Condensate
5. Solids (produced sand and debris)
6. Chemical inhibitors

Critical problems associated with these issues include:

1. Flow along and across the tieback is accompanied by rapid and transient pressure and temperature, composition, and pattern changes that need to be monitored for effective management.
2. Presence of gas and free water in the pipeline may result in hydrate formation, which needs to be managed. A typical strategy is the use of hydrate inhibition chemicals such as mono-ethylene-glycol (MEG) whose molecular weight/density is different from those of other phases. The hydrates are formed at certain pressure and temperature-operating windows.
3. Presence of incompatible water with different salinities will result in scale formation. There is a scale formation envelope. Likewise, for alsphatene or wax deposition.
4. Solids production from the wells and nucleation of hydrate and scale crystals accompanied by transport and deposition may cause blockage and

potential erosion of line chokes/jewels. The production of sand and other solids into sub-sea tiebacks and manifolds can cause serious problems such as reduction in the available space open to flowing fluid; additionally, the sand can cause blockage and/or erosion of piping, fittings, and valves. The potential direct impacts of these are undesirable pressure drop, loss of production, and possible safety hazards.

2.10.1 Scale Management

Scales are deposits that form on and are adherent to a surface, usually a metal surface. Mineral scales found in oil production systems contain inorganic salts such as the sulphates and carbonates of calcium, strontium, and barium. Other common constituents of scale deposits are iron-oxides and hydroxides, iron-sulphide, and evaporite salts. Waxes and asphaltic materials are commonly entrained in deposits and can bind together friable solids. Sludges can be described as accumulations of unconsolidated material; for example, scale fragments, sand, waxes, and corrosion products that do not strongly adhere to surfaces. Sludge deposits collect in zones of low flow or at bends, low points, storage tanks, etc.

The deposition of scales and sludge in oil production equipment or within the reservoir is a common, widespread, and potentially serious problem. The deposition of scale deposits on tubular surfaces reduces the fluid flow by reducing cross-sectional area and increasing surface roughness. Fluid flow within reservoirs is restricted due to blockage of pores. On heat exchanger surfaces reduced heat transfer properties cause declines in efficiency. One localised 'advantage' of scale deposition on steels is reduced corrosivity due to the formation of a thin passive of mineral scale. However, bio-fouling can occur if stagnant brine persists beneath scale deposits.

2.10.1.1 Formation and Prediction of Mineral Scales

Oilfield brines contain high levels of dissolved salts and gases. Within the reservoir there is equilibrium but when this equilibrium is disturbed in the production process mineral deposits will occur. The formation of scales from flowing brine solutions occurs in stages. In order for a mineral salt to form a hard, consolidated scale deposit it is generally accepted that three conditions have to be satisfied:

- Brine has be supersaturated with respect to the scaling minerals
- Adequate nucleation sites must be available for crystal growth
- Sufficient contact time to enable the growth of a consolidated deposit

The different Scales are:

1. Calcium Carbonate: Formation and Prediction

Calcium carbonate scale, lime scale or hardness scale is widespread within water handling processes in all manufacturing industries and often appears as clear or translucent crystals, discoloured by impurities (eg, orange - red by iron). Deposits are recognised by being highly crystalline (in most cases) and by their reaction with hydrochloric acid: a strong effervescence occurring as CO2 is formed (other carbonates give this reaction so it is wise to check for soluble calcium). Oil fouled scale surfaces may show little or no reaction to acid because the acid cannot wet the mineral surface.

The solubility of calcium carbonate in water is very low, e.g., below 1 gram per litre at 20 °C under 1 atmosphere CO_2 pressure. Its solubility is strongly influenced by partial pressure of CO_2, temperature, and concentration of other salts in solution. Changes in super saturation and hence precipitation and scaling usually occur as a result of temperature increase or pressure decrease (drawdown). Calcium carbonate is often the first scale type to deposit. It is characteristic of produced oilfield brines and is very commonly found at the onset of first water production, and typically severe at points of large pressure drop such as at the production zone and at chokes. In solution, calcium and bicarbonate ions (not carbonate) are at equilibrium with dissolved carbon dioxide and dissolved calcium carbonate:

$$\underset{\text{(in solution)}}{Ca(HCO_3)_2} \rightarrow H_2O + CO_2 + \underset{\text{(scale)}}{CaCO_3}$$

In order to predict the conditions under which scale-forming minerals may precipitate it is necessary to know how the solubility product of the mineral varies with changes in brine composition, temperature, and pressure.

2. *Barium and Strontium Sulphate: Formation and Prediction*

Barium and strontium Sulphate-Scale deposition is usually the result of two different brines co-mingling: one-brine rich in sulphate ions and the other rich in alkaline earth ions. When this happens, the solubility product of the mineral sulphate is exceeded and the water becomes supersaturated. This process is generally responsible for the formation of barium and strontium sulphates scales:

$$\begin{array}{ccccc} Ba^{2+} & + & SO_4^{2-} & \rightarrow & BaSO_4(\text{solid}) \\ Sr^{2+} & + & SO_4^{2-} & \rightarrow & SrSO_4(\text{solid}) \\ (\text{from solution A}) & & (\text{from solution B}) & & \text{scale} \end{array}$$

Barium and strontium sulphate scales are typically crystalline but opaque: the crystal size usually being smaller than is typical for calcium carbonate. Colour may be pale buff to deep brown/red if contaminated by iron minerals. Acids do not attack sulphate scales. Strontium readily exchanges for barium in the crystal lattice, and deposits containing both barium and strontium are common. The solubility of barium sulphate is about 0.002 gram/litre in distilled water at 25 °C, which is about two orders of magnitude less than strontium sulphate and also significantly less than calcium carbonate. Co-mingling of injection and formation waters is the primary cause of precipitation

TABLE 2.2 Typical Ion Analysis of North Sea Formation Water and Seawater

Formation Water	Ion	Seawater
29,370	Na	11,020
370	K	450
500	Mg	1,400
2,808	Ca	420
575	Sr	6
252	Ba	0
0	SO_4	2,750
490	HCO_3	140
52,350	Cl	19,700

As the two waters (above) mix virtually all barium and a proportion of the strontium will be removed as the sulphate salts. Barium and strontium sulphate solubility increases to a maximum as ionic strength is increased. The tendency for scale to form is affected relatively little by temperature and pressure. A comparison of the solubility of barium and strontium sulphate at different temperatures is given in Tables 2.3 and 2.4.

Solubility data been used as the basis of predictive models. As barium and strontium sulphate precipitation is not strongly influenced by temperature or pressure, predictive models are generally considered to be more reliable than those for calcium carbonate. The output from the model is usually expressed as mineral saturation index plotted as a function of the proportion of each water component present. An example of such a plot is given in Figure 2.22, which shows the results for both types of calculation for mixing two waters (formation water and seawater).

3. *Calcium Sulphate: Formation and Prediction*

In many oil-producing areas, gypsum ($CaSO_4$. $2H_2O$) is formed as a result of the commingling of incompatible brines from different producing zones. It is an uncommon scale in the North Sea. Calcium sulphate can also be formed as a result of significant temperature or pressure fluctuations. Its water solubility is higher than that of barium or strontium sulphates. The unplanned loss of calcium chloride completion/workover brines to reservoirs may also lead to unexpected calcium-sulphate deposits. The solubility of gypsum is higher than either barium or strontium sulphate and decreases with temperature above

TABLE 2.3 Solubility of Barium Sulphate at Different Temperatures and Sodium Chloride Levels

% NaCl	25 °C	50 °C	80 °C
0	2-3	3-4	4-5
1	12	15	20
3	17	24	30
10	28	37	55

TABLE 2.4 Solubility of Strontium Sulphate at Different Temperatures and Sodium Chloride Levels

% NaCl	25 °C	95 °C
0	132	113
10	630	691

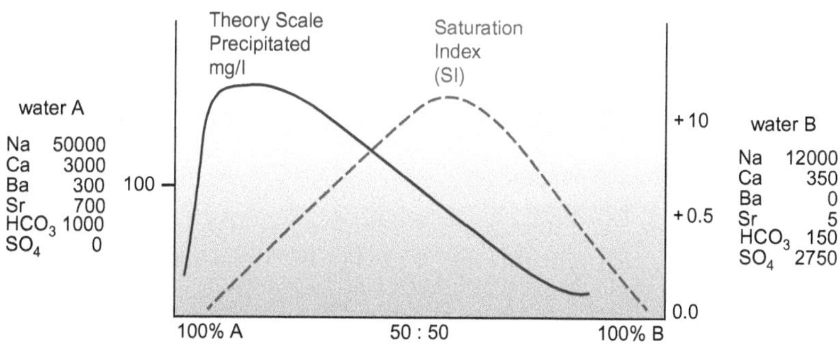

FIGURE 2.22 Scale Prediction Output Data.

about 40 °C. Magnesium ions have a significant effect on solubility. Approximately 2-gram/l calcium sulphate can theoretically be held in distilled water solution at 20 °C. Other mineral forms of calcium sulphate may occur, and more advanced prediction models are able to provide this data. However, prediction models are only as good as the data that is fed to them.

4. *Iron Scales*

Iron scales can result from the accumulation of corrosion products (iron-oxides, hydroxides) or can be derived from the interaction of soluble iron with its environment. Formation water commonly contains 1-10 ppm soluble iron and occasionally up to 60 ppm. Iron-sulphide is deposited whenever fluids containing H_2S contact carbon steels. Bacteria play a role in the conversion of ferrous ions into the poorly soluble ferric hydroxide and sulphides. Ferric sulphide and ferric hydroxide form gelatinous deposits that can very rapidly plug formation pores. Iron scales vary considerably in appearance but are generally dark coloured. Water containing colloidal ferric oxide is red in colour and water containing ferrous-sulphide is black.

5. *Silica Scales*

Oilfield brines contain low levels of silica (typically < 25 mg/litre). Higher levels than this may be associated with hot/deep wells, but it is uncommon for silica to be a major problem in oil flowlines. Precipitation of silica can occur wherever there is significant evaporation of steam as in high-pressure boilers and on heat exchangers (geothermal wells have significant silica problems). Silica scales are very hard, difficult to inhibit, and difficult to remove.

2.10.1.2 *Layered Scale Deposits*

Deposits removed from tubulars and separators are frequently layered. Each individual layer represents a sequential period of scale deposition and may be mineralogically different from the previous layer. Dark-coloured bands are usually caused by iron or wax contamination.

2.10.1.3 *Scale Prevention*

When there are conditions for scales to form there are three options.
1. Leave the solids in and allow the scale to form and treat periodically

- Mechanical removal methods
- Chemical scale removers

2. Pre-treat brines that will mix by removing the dissolved and suspended solids

- Ion exchange resin to soften water
- Reverse osmosis
- Nano-filtration

3. Leave the dissolved solids in and treat to prevent deposition

- Add scale inhibitors
- Add scale dispersant
- Use magnetic devices
- Use electronic devices

The periodic removal of formed scale as the only treatment strategy is possible if:

- Scale is carbonate rich, accessible, and dissolved by acids
- Scale formation is low and predictable
- Downtime does not represent a major cost penalty

Scale prevention is much more cost effective and practical than scale removal. Scale removal will be discussed in a later section. Scaling conditions can be avoided by removing one or more of the scale-forming ions from solution. Reverse osmosis removes all organics and up to 99% of all ions but is not viable for very high dissolved ion content brines. Nano-filtration is similar to reverse osmosis but only removes divalent and trivalent ions. Specific membranes can remove over 95% of sulphate ions so will reduce the sulphate concentration of seawater from 2,700 ppm to 80 ppm so it can be injected into high barium containing formation brines. This type of treatment is efficient but expensive.

The more usual treatment option selected is chemical scale inhibition. The use of inhibitors is relatively inexpensive, efficient, and can be applied to the processing system at the appropriate point, be it the reservoir, tubulars, topside process, or water injection.

2.10.1.4 Scale Inhibition Processes

Scale inhibitors are broadly classified as threshold inhibitors (crystal growth modifiers, nucleation inhibitors), dispersants, and complexing agents. Scale inhibitors act by interfering with the processes of nucleation and/or by restricting crystal growth. Nucleation inhibitors must be able to diffuse readily in the bulk brine but must be sufficiently large to be able to disrupt formation of ion clusters. Thus, a balance between acceptable rate of diffusion and effective size (conformation) is necessary. Crystal growth occurs on active sites (such a steps) on a crystal face. Good crystal growth inhibitors should have a strong affinity for these active sites and should be able to diffuse to other sites on the surfaces of the crystal. The presence of an inhibitor molecule at the crystal surface is energetically more favoured than the same molecule in bulk solution, especially if the inhibitor has many potential bonding groups. As the active sites only represent a small percentage of the exposed scale crystal surface, inhibitors can be effective at very low doses. Most commercial products are able to operate by both nucleation prevention and crystal growth restriction. Scale inhibitors can be quite specific in their action toward individual scaling minerals. The distance between the important coordinating atoms in the scale inhibitor molecules should match those on one of the crystal faces for high efficiency.

Figure 2.23 shows how surface-active molecules can modify crystal growth by the reversible adsorption of anionic and cationic surfactants to adipic acid.

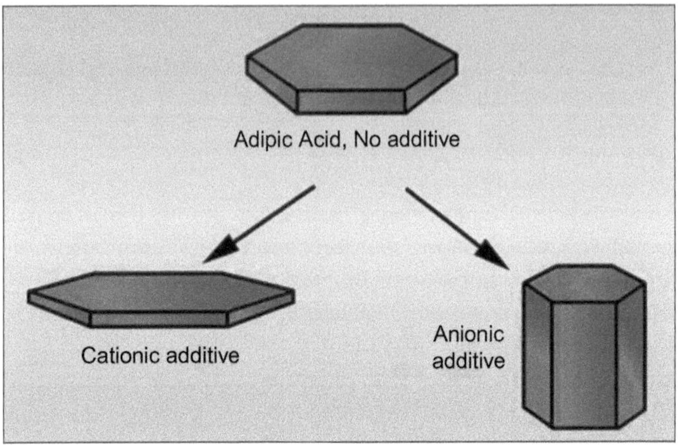

FIGURE 2.23 Crystal Inhibition by Crystal Habit Modification.

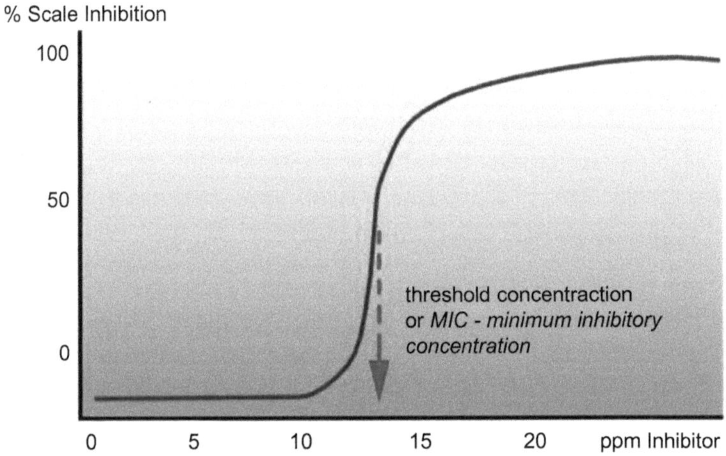

FIGURE 2.24 Performance Characteristics of Threshold Inhibitors.

By interfering with the process of growth the rate of scale mineral growth is very much reduced. As the adsorption is reversible, extremely low, substoichiometric concentrations of inhibitors, typically 2-20 ppm, are required and the term 'threshold inhibition' is applied. Typically, the performance of threshold inhibitors increases suddenly above a specific concentration (the threshold or minimum inhibitory concentration, MIC). A typical curve for concentration of inhibitor against the effectiveness is illustrated in Figure 2.24. Polycarboxylates and phosphonates are common threshold inhibitors and work either by modifying crystal growth or by reducing the rate of nucleation (both kinetic effects). The two general classes of threshold scale inhibitors that are in common use in oilfield production systems are phosphorous-containing compounds such as phosphates and phosphonates and polycarboxylates.

2.10.1.5 Examples of Scale Inhibitors

1. **Phosphonate Inhibitors:** Metaphosphates, including sodium hexametaphosphate and polyphosphates, are very effective against calcium-carbonate and calcium-sulphate scaling, and they are the least expensive class of scale inhibitor. They were the first inhibitors to be used in commercial quantities and are deployed in water-treatment plants in many industries. Their use in oilfield water-flooding operations dates back to at least 1940 but is no longer commonplace. The thermal stability of dimeric meta-phosphate inhibitors is very poor. The maximum recommended application temperature is about 70 °C.

   Phosphate-containing inhibitors are believed to function mainly by interfering with the nucleation process. Organic phosphate esters (such as the hydroxylamine-phosphate ester type) are slightly more stable than polyphosphates but are still too sensitive to high temperature and low pH conditions to have wide application. They are effective against calcium carbonate and calcium sulphate and show excellent brine compatibility. They have found limited application in oilfield systems. At pH 7 and above triethanolaminephosphate ester is effective against barium sulphate. Aminomethylene phosphonates (AMP and homologues) provide high thermal stability and have found much wider application in oilfield applications than other phosphorous-containing types and can be used in systems up to about 120 °C.

   All commercially available phosphonates are effective against calcium-carbonate scale. Diethylenetriaminepentamethylene phosphonate (DETA phosphonate) is effective against carbonates, barium, and strontium sulphate and can be used for 'squeeze' applications. This product has been used extensively in the North Sea. A phosphonomethylated polyamine is available and is claimed to have better squeeze performance than other phosphonates.

   Hydroxybisethylidene phosphonic acid (HEDP) is a non-amine based phosphonate inhibitor. It has some advantages when used topside where the chloride ion content is low, excellent carbonate inhibition, high brine compatibility, good iron sequestration properties, and thermal stability. However, it is ineffective against barium sulphate.
2. **Polycarboxylate Inhibitors:** The use of organic polymers as scale inhibitors has been recognised for a long time with early patent literature advising the use of natural products such as starches, gums, lignins, tannins, and even potato extract! The most important commercial products are based on polyacrylates, polymaleates, and polyvinylsulphonates and various co-polymers and terpolymers. These are classed as dispersants.

Some polymer inhibitors, particularly the higher molecular weight acrylate based products, are sensitive to soluble calcium, which in sufficient concentration can precipitate the inhibitor as a calcium salt. Users should generate solubility 'phase maps' to describe the tendency of inhibitors to precipitate under various conditions of pH, temperature, inhibitor content, and calcium concentration. This technique is essential when designing inhibitor treatments for reservoir squeezing. Polymer inhibitors are preferred if the major problem is sulphate scaling and especially if the temperature is relatively high. One major advantage of polymer types is that thermal stability is excellent: they are therefore suitable for application in hot, deep wells. Polymeric inhibitors are also generally less corrosive in concentrate form than most of the phosphorus-containing types.

Polyacrylates and co-polymers have been used extensively for 'squeeze' applications in many oil provinces, providing long-term performance and thermal stability. Polyaspartic acid products have recently been shown to function as inhibitors of both scale and corrosion and, unlike other scale inhibitors, are completely biodegradable.

3. **Oil-Soluble Scale Inhibitors:** Oil-soluble scale (squeeze) inhibitors were introduced in 1999. They are used primarily in situations where it is undesirable to place a significant amount of water into a well or if it is desirable to squeeze a well soon after completion and before first water production (protect against early scaling). The inhibitor molecule partitions to the water phase.
4. **Solid Inhibitors:** Slow release of inhibitor can be achieved if the product is formulated in a solid form. When placed in the scaling environment the solid material slowly dissolves and releases the active inhibitor to the aqueous phase.

   Two alternative technologies were developed during the mid-1990s for oilfield applications. The first involves the micro-encapsulation of active scale inhibitor (e.g., phosphonate) into a slow-dissolving polymer matrix with a closed cell structure. The capsule also contains sufficient weighting agent to ensure a density higher than most oilfield brines. These capsules are about 2-3 mm in diameter and can be slurried in water and pumped to the well bottom. The normal placement would be to the rat-hole. Other technology uses ceramic beads impregnated with scale inhibitor.

2.10.1.6 Evaluation of Scale Inhibitors

The suitability of a scale inhibitor for a specific application can be evaluated in the laboratory. Performance tests that are performed regularly include performance by static jar test and performance by dynamic tube-blocking.

1. Static Jar Test

The static jar test measures the ability of candidate scale inhibitors to prevent the precipitation of specific mineral salts within glass jars. This is a valuable screening test for scale inhibitor evaluation and ranking of performance.

Tube-Blocking Tests

Unlike the static jar test that measures static precipitation, the tube-blocking test actually measures scale deposition in a capillary under dynamic conditions. The test is equally suited for the evaluation of carbonate and sulphate scaling, so is useful for measuring the total scaling condition. It is a much more reliable screening method for testing inhibition of calcium carbonate than the jar test.

2. Brine Compatibility

Scale inhibitors destined for continuous application at the wellhead, for example, should be fully compatible with the various produced brines that may be contacted. This is tested by adding the scale inhibitor, as supplied, to the produced brine at use concentration and at higher levels to simulate overtreatment. A turbidity or precipitate, which can be the calcium salt of the inhibitor, represents a failure.

Scale inhibitor compatibility is pH and temperature dependent, and chemical companies exploit this feature in order to design treatments for which a controlled amount of precipitation is allowed in the reservoir to enhance performance longevity, called a 'precipitation squeeze'.

3. Compatibility with Materials of Construction and Other Chemicals

Corrosion risks from polycarboxylate or phosphonate inhibitors at use concentrations are generally minimal. Acid phosphonates can contain high levels of chloride ion and care must be taken during storage and transfer of concentrated solutions to minimise corrosive effects to pipe work, transfer pumps, etc. Unless the inhibitor is formulated in a glycol derivative or an alcohol solvent it is unlikely that there would be compatibility problems with elastomers or plastics. Scale inhibitors are anionic and corrosion inhibitors are cationic. Compatibility problems can occur if significant levels of both types of inhibitor are present in the same brine. Phosphonates are incompatible with some biocides.

4. Reservoir Compatibility

A 'core flood test' will determine if an inhibitor will cause damage to a given reservoir rock. The same test can be used to evaluate adsorption behaviour. Selecting suitable reservoir material and the preparation of representative core plugs is a highly skilled operation. To evaluate reservoir compatibility, the various treatment solutions are pumped through a representative core plug in a pressurised vessel and the effects these solutions have on relative permeability are monitored. By collecting effluent samples and analysing for inhibitor content it may be possible to make judgments about the adsorption and desorption processes. This type of data is used to compare two inhibitors or inhibitor treatments.

5. Thermal Stability

Two aspects are important: the stability in storage and the stability in use. Storage in warm countries for extended periods can be problematic (degradation, etc.). Storage in cold climates at temperatures below the freezing point of water can cause the product to become too viscous to pump or, worse still, to freeze. Prolonged storage at low temperatures can result in phase separation.

6. Environmental Impact

Scale inhibitors are not normally thought to have a significant long-term impact on the environment and are generally of a low order of toxicity, although in common with other surfactants they are not readily biodegraded. Inhibitors are water-soluble and so will be discharged along with the process water. If this discharge is into the sea then very rapid dispersal will occur.

2.10.1.7 Scale Inhibitor Application

The two application techniques considered here are continuous application to flowlines and squeeze treatment.

1. Continuous Application

From jar tests and tube-blocking experiments, suitable scale inhibitors will have been identified and minimum threshold levels determined. It is operationally advisable to apply the inhibitor as far upstream of the problem area as possible. In production wells this usually means in the wellhead area. Down-hole continuous application is possible via narrow bore treatment strings or gas lift, although maintenance and installation costs are possible disadvantages.

It is essential that the equipment chosen to apply the scale inhibitor is positioned properly and is in the correct location to ensure rapid dispersal into the fluids. Scale inhibitors are formulated in a water base to ensure rapid dispersal of the active ingredient. The product should be compatible with materials used in the dosing pumps and lines and be compatible with the brine and any other production chemical added at the wellhead. Demulsifiers and corrosion inhibitors are also usually added at the wellhead and ideally the points of addition should be separated by a few metres. Dosage levels to production lines are normally in the order of 5-25 mg /litres based on water production, although in severe brine conditions up to 200 mg/litre have been known to be necessary. The applied dose should be slightly higher than the threshold value determined by tube-blocking experiments.

Scale inhibitor is also dosed continuously to water injection systems, although the dosage requirement is much reduced, typically between 1-5 mg/litre. Often, the deployment of inhibitor to injection systems is stopped after a pre-determined injection volume.

2. Down-hole Squeeze Application

To protect the wellbore, perforations, bottom-hole area, and tubulars an alternative deployment method is necessary. The squeeze technique is commonly used and involves the placement of a concentrated slug of scale inhibitor into the formation by 'bull heading' a concentrated slug down the production tubular. The inhibitor is displaced into the formation to a predetermined distance from the wellbore by overflushing with neutral brine. After a soak period in which the inhibitor adsorbs or precipitates, the well is returned to production. The success of the treatment depends on the rate of release of the inhibitor back to the produced water where a slow but steady return profile is most desirable.

3. Adsorption and Precipitation Squeeze Treatments

Inhibitor is retained in the reservoir either by adsorption to rock surfaces or by precipitation. In almost all cases both adsorption and precipitation mechanisms will occur together to some extent, regardless of the actual design. Scale inhibitor is squeezed as a concentrate (5- 15% active), typically at pH 3-6, and is retained in reservoir about 3 m from the wellbore for up to 24 hours. The well is returned to production and the inhibitor desorbs slowly to the aqueous phase over a period of months.

The squeeze life is the period during which there is sufficient inhibitor to protect the well from scale formation. This period varies according to the strength of inhibitor, the volume injected, the distance from the wellbore, and reservoir characteristics. However, there is a finite adsorption capacity within a given zone of reservoir rock and simply increasing the amount of inhibitor is not guaranteed to be successful. A major disadvantage of the adsorption squeeze is that typically 50% of the product is irreversibly lost to the formation or is not adsorbed in the near wellbore and flows back during the first few hours of production. Retention may be enhanced by precipitating a proportion of the inhibitor as a sparingly soluble salt that is available for slow solubilisation back to the production system.

Calcium chloride can be used as the precipitant and may be mixed with the inhibitor or introduced in a preflush or afterflush solution. By pumping the inhibitor plus calcium chloride solutions in mildly acidic form, it is possible to rely on the increased temperature or higher pH conditions in the reservoir to induce precipitation. Calcium phosphonate or calcium polycarboxylate salts are sparingly soluble and their solubility decreases with increasing temperature and increasing pH.

4. Monitoring of Produced Water

Once a well has cut water, ion tracking is recommended to monitor the scaling potential (prediction) and also to monitor the performance after squeeze treatment. Regular 12-ion assay (cations and anions) plus pH and scale inhibitor analysis on the produced water from each well will provide the basis for interpretation of well performance. Multivariate analysis is a useful tool that will help identify small perturbations in analytical parameters and may indicate optimum time for squeezing (and resqueezing). Chromatographic techniques are available for determining scale inhibitor residuals but non-phosphorus containing products are subject to significant error at 5 mg/litre and below.

2.10.1.8 Scale Removal

Non-chemical options to remove existing scale deposits include reaming (drilling), scraping, and water jetting. Non-chemical methods can be the only viable option for treating thicker, consolidated deposits. In general, if the scale is accessible and thin enough to allow good surface contact then the use of chemicals is usually cost-effective. Scales must be removed to maintain throughput of fluids and to maintain pumps and valves in good working order. Failure to do this can compromise safety.

2.10.1.9 Chemicals Used in Scale Removal Operations

Design of a scale removal treatment should be considered in stages:

- Estimate amount of scale from volumes of water produced, scale prediction models, and calliper surveys and calculate volumes of chemical treatments
- De-grease to allow mineral surfaces to be water-wet and so susceptible to attack from chemicals
- Major scale remover treatment to remove bulk of deposit
- Post-treatment to minimise risk of corrosion or re-fouling of surface

Pre-treatment (surfactants and degreasers) and post-treatment (washing, passivation) are not always essential to the success of a scale removal treatment process but the consequences of not using these additives should be considered.

1. Degreasers as Pre-Treatment

Surfactants or mutual solvents such as butyl-glycol ethers are suitable as down-hole pre-treatment fluids as they can displace or solubilise oils from the scale surfaces. Strongly charged surfactants should be avoided as they may affect the wetting character of reservoir rock or cause foam problems.

Heavier oils, waxes, and asphalts can also be removed by pre-treatment with aromatic solvents. Bacterial films can be particularly difficult to remove but oxidising biocides or slug doses of gluteraldehyde may be effective.

For down-hole applications the degreaser/surfactant may be pumped in advance of the acid or incorporated in the acid solution itself to simplify the overall treatment process.

2. Inorganic Acids

Hydrochloric acid (HCl) is very widely used and is inexpensive. It reacts readily with calcium-carbonate scale deposits, has a high capacity for dissolution, and results in soluble reaction products. The acid is typically sold in strengths of 5-15%. Reaction with all clean carbonate minerals is vigorous, producing carbon dioxide gas.

HCl is very corrosive to steel, including high chrome steels such as Duplex. Acid strengths in excess of 15% are not normally recommended due to the severe corrosivity. Corrosion inhibitors should be added (0.05-0.5%) to reduce the corrosion rate to an acceptable level. Commercial 'inhibited' hydrochloric acid is available. Inhibitors only protect against corrosion at temperatures below about 70 °C.

Blends of hydrochloric acid and mutual solvents (e.g., glycol ethers) together with corrosion inhibitors and surfactants are offered by chemical companies: these proprietary products can effectively penetrate hydrocarbon layers enabling the acid to contact the scale deposit.

Secondary precipitation fouling is frequently associated with the use of hydrochloric acid for down-hole scale dissolver treatments. HCl dissolves calcium carbonate and also reacts with iron compounds siderite (ferrous carbonate), pyrite (ferrous-sulphide), and iron-oxides magnetite (Fe_3O_4) and haematite (Fe_2O_3) at a similar rate but as the acid spends on the matrix and the pH rises, solutions containing iron salts are susceptible to re-precipitation and reservoir plugging can occur:

- Above pH 2.2, Fe^{3+} (in solution) $\rightarrow$ $Fe(OH)_3$ (insoluble)
- Above pH 7, Fe^{2+} (in solution) $\rightarrow$ $Fe(OH)_2$ (insoluble)

Care must be taken if the system to be cleaned contains iron-sulphides as these react with hydrochloric acid producing the toxic gas hydrogen-sulphide. More details are given in a SPE monograph [20].

Nitric acid (HNO_3) is useful for removing tight adherent oxide films from stainless steel and high nickel alloys (pickling action). It is rarely used for the removal of common oilfield scales. Nitric acid is very corrosive to mild steel and copper alloys and is dangerous to handle because it is a powerful oxidant and produces toxic fumes.

Sulphamic acid (NH_2SO_2OH), often referred to as powdered acid, is frequently used as a replacement for hydrochloric acid even though it is more expensive and reacts slower with carbonate scales. Its advantages include its availability as a dry powder (ease of transportation, safe to handle), non-volatility, non-hydroscopic, odourless, less corrosive than HCl, and does not give off acid fumes in storage and deployment. Phosphoric acid (H_3PO_4) is sometimes used for the removal of iron scales, particularly iron-oxide films, and it is the basis of many 'domestic' rust removers. Its use in the oil industry is limited by the risks associated with the precipitation of calcium phosphate.

3. Organic Acids

Citric acid and its ammonium salts are useful for the removal of rust and iron-rich scales and is frequently used in situations demanding very low corrosivity. The acid forms stable complexes with both ferric and ferrous iron up to pH about 10 and therefore the risks of re-precipitation of iron hydroxide are minimal. Addition of citric acid to other acid cleaning formulations is often recommended to provide iron-sequestering properties to the formulation.

Formic acid (HCOOH) and acetic acid (CH_3COOH) have found significant use as effective dissolvers for calcium-carbonate deposits. Although not so strong as hydrochloric acid, they are used for down-hole treatments and in situations where a relatively fast reaction is required but with minimal corrosivity as these acids are less corrosive than hydrochloric acid. Both acids are used at 5-15% strength. Higher strengths are not normally recommended as the calcium salt has a limited solubility that, once reached, inhibits further reaction. The organic acids can be used to higher temperatures than hydrochloric acid but similarly need corrosion inhibitor for oilfield application.

4. Sequestering Agents and Sulphate-Scale Removal

Sulphate scales are not dissolved by mineral or organic acids [21]. Polyaminocarboxylates, such as ethylenediamine tetraacetic acid (EDTA) and nitriloacetic acid (NTA), are sequestrants and widely used as their neutral to alkaline salts. They can be effective for the treatment of sulphate scales including barium and strontium sulphates, and there is a significant amount of these salts sold to oilfield markets.

Diethylenetriamine pentaacetic acid (DTPA) is particularly favoured for barium sulphate and is used as the sodium salt, but improved dissolution can be obtained under certain conditions by using the potassium or ammonium salts.

5. Converters

Calcium sulphate (gypsum) can be successfully treated by strong alkalis resulting in the 'conversion' of the hard sulphate scale to another solid form that is more easily removed by dilute acid. Sodium and potassium hydroxide, sodium carbonate, ammonium carbonate, and other compounds have been used as alkali. The reaction is:

$$CaSO_4(\text{solid}) + NaCO_3 \rightarrow CaCO_3(\text{solid}) + HCL \rightarrow CaCl_2(\text{soluble})$$

The calcium carbonate will now react with hydrochloric acid. The technology is well established, especially for onshore, low-producing wells.

2.10.1.10 Radioactivity in Oilfield Scales

Radioactive species from the uranium-238 decay chain are naturally present in reservoirs that contain oil, gas, and water. Radium-226 is one of the longer-lived daughters and therefore can persist and equilibrate: as it is chemically related to barium and strontium it is usually found in the connate brine. Radium decays to radon, an inert gas that can migrate through the rock structures. Most of the members of the decay chain below radon are short-lived and typically occur in produced fluids associated with the hydrocarbon phase. Lead-210 is the other most significant isotope of this decay chain.

Barium Sulphate-Scale deposits frequently contain low concentrations of (coprecipitated) radium sulphate. These low specific activity (LSA, or NORM) scales are problematic, as they can be sufficiently radioactive to be covered by regulation regarding their safe disposal. Mechanical methods of scale removal may result in the generation of toxic radioactive dust.

2.10.2 Hydrate Nucleation and Management

Under closely defined conditions of high pressure and low temperature low molecular weight hydrocarbon gases, including several of the components of natural gas, will react with water to form crystalline clathrate-like compounds, or gas hydrates. Figure 2.25 gives the three-dimensional structure of these solid forms of gas.

Hydrates have become a major problem as they can block transmission lines, plug blowout preventers, cause tubing and casing failure and foul pipes, heat exchangers, and valves. Operators of multiphase transport pipelines and other sub-sea production are particularly vulnerable to hydrate production as high water cuts and low temperatures are expected for an appreciable part of the field life. In one situation in the mid-1990s, a 600-metre-long solid hydrate plug formed in a 10-inch sub-sea riser line: several weeks circulation using ethylene glycol in coiled tubing was needed to free the line. It should be remembered that hydrates do act to concentrate lower hydrocarbons and large natural reserves exist in deepwater and in permafrost (1 cubic metre of hydrate can hold nearly 2000 scf of gas).

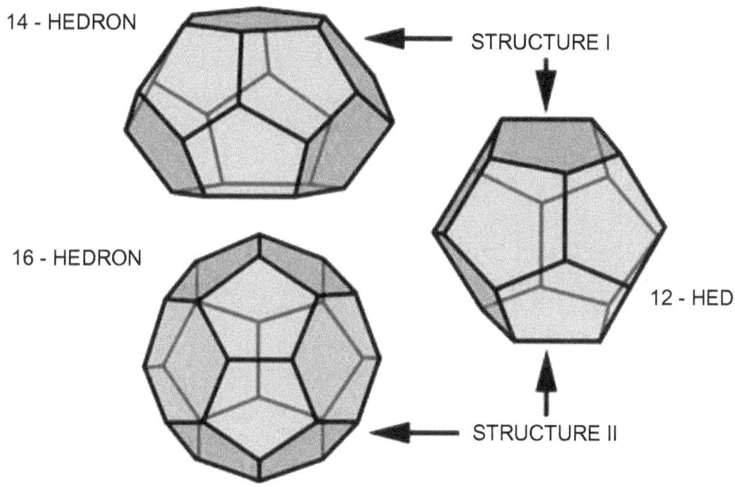

FIGURE 2.25 Structure of "Ice" Formed from Water and Methane or Ethane.

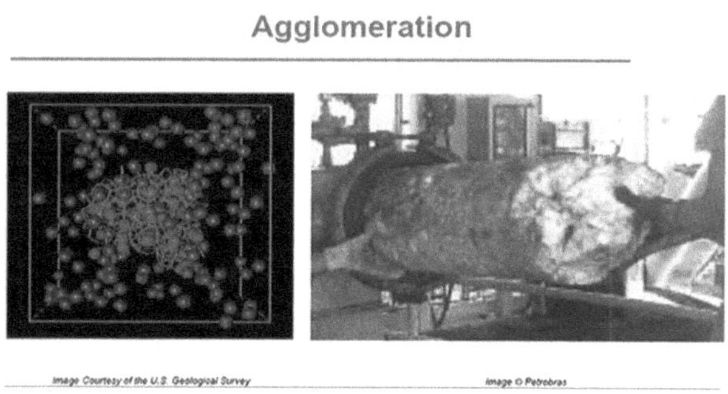

FIGURE 2.26 Hydrate Agglomeration.

2.10.2.1 Hydrate Formation

Under critical conditions of temperature and pressure water and low molecular weight hydrocarbons such as methane, ethane, propane, and butane are encapsulated within an aqueous hydrogen-bonded structure. The material grows by encapsulating more and more water. The conditions for hydrate formation for methane and water are given in Figure 2.27. This shows that conditions of high pressure and low temperature favour hydrate formation.

Hydrates can form as a simple consequence of temperature decline, which in turn can be the result of local pressure, drops due to turbulence, or gas expansion. The consequence is that even at normal temperatures, it is possible for solid hydrates to continue to grow so that eventually the deposits completely block lines. Prediction of hydrate formation is possible. Knowledge of the hydrocarbon composition, the brine content and salinity, and of the temperatures and pressure profile of the system enables calculation of the risk of hydrate formation and enabling phase diagrams to be constructed showing the hydrate region. Mathematical models have now been developed (e.g., HYSIM/PROVISION) and may include parameters for the common thermodynamic inhibitors, methanol, ethylene glycol, and di- and tri-ethylene glycols.

Salt has an inhibiting effect on hydrate formation (lower gas solubility) so the risks associated with hydrate formation are reduced in high TDS-produced brines.

2.10.2.2 Prevention Options

It is possible to prevent hydrate formation by:

- removing one of the components – water or hydrocarbon;
- increasing temperature or decreasing pressure; or
- adding a chemical inhibitor.

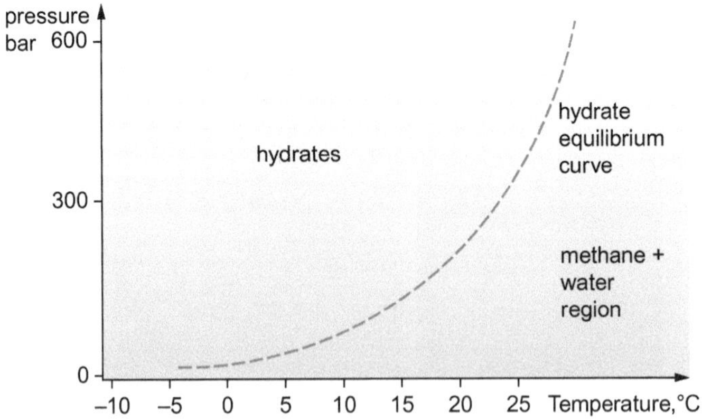

FIGURE 2.27 Pressure–Temperature Curve for Hydrate Formation in Methane–Water System.

Down-hole separation and re-injection of water would remove a proportion of water reducing risks. Tri-ethylene glycol is used in gas treatment to dry gas. This effect is shown on Figure 2.28 where the curves for hydrate treatment with and without ethylene glycol are given. The treatment moves the curve to the right so that no hydrates will form at higher pressures and lower temperatures.

Pipelines can be insulated or buried so that internal temperatures are high enough to ensure the fluids remain outside the hydrate region during normal flow conditions. Gas hydrate formation may be controlled by the application of chemical inhibitors. There are two distinct types of inhibitor: traditional thermodynamic solvating types such as glycol and methanol and the so-called threshold type inhibitors that include kinetic and anti-agglomerators.

2.10.2.3 Hydrate Chemical Inhibitors (Figure 2.28)

Thermodynamic Inhibitors Alcohols and glycols change the solvating properties of the aqueous phase. They readily form hydrogen bonds with water molecules making gas entrainment more difficult. As a result, the hydrate equilibrium curve is permanently displaced to lower temperatures or to higher pressures. Methanol and ethylene glycol are the main products deployed but in order to be effective must be added continuously at concentrations typically 10-50% based on the aqueous phase. Alternatively the product can be added as a slug dose to pipelines/wells prior to shut-in periods.

The advantages can be identified as:

- good track record;
- easily pumped through pipelines;
- no fouling at point of injection; and
- not dependent on oil composition.

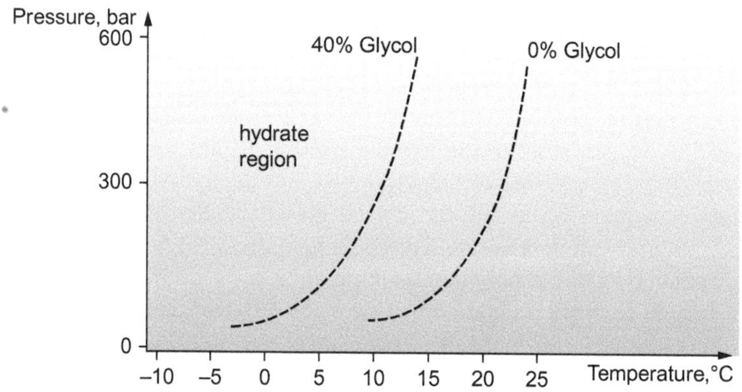

FIGURE 2.28 Pressure–Temperature Curve for Hydrate Formation in Methane–Water System Treated with Ethylene Glycol.

Note: Although partial recovery of methanol or ethylene glycol is often possible, the costs of these products become very significant when the large volumes deployed and transport and storage logistics are considered. Other disadvantages include toxicity fears and high flammability of methanol. Di- and tri-ethylene glycol (TEG) are also used in gas drying. The removal of water from a gas stream will significantly reduce any risk of hydrate formation downstream. Considerable volumes of these chemicals are sold to the oil production industry.

Threshold Hydrate Inhibitors (THI) Chemical inhibitors that act at much lower concentrations than thermodynamic inhibitors by delaying or reducing the rate of hydrate formation are called threshold hydrate inhibitors (THI). They are sub-divided into two types:

- Kinetic inhibitors (KI):
  - Complex polymers with mixed water/oil solubility
  - N-vynyl-2-pyrrolidone polymers and co-polymers
  - Terpolymers
- Anti-agglomerator types:
  - Surfactants that prevent or delay agglomeration of hydrate crystals
  - Quaternary compounds

Kinetic hydrate inhibitors (KHI) are products that delay the onset of autocatalytic hydrate formation and are called kinetic inhibitors. Mechanistic explanations have been explored by Lakvam and Ruoff [(1)]. Performance is characterised by an induction period during which there is little or no uptake of gas into the water phase. The induction period is followed by rapid hydrate formation, similar to the situation in the absence of inhibitor. The overall effect for a continuously dose product is that the hydrate equilibrium curve is shifted to lower temperatures. It has been found that doses of 0.1-1% inhibitor can give induction periods of in excess of 24 hours compared to a control of essentially zero induction time. Co-polymers of N-vinyl 2-pyrrolidone and methylacrylic or co-polymers of vinyl pyrrolidone and vinyl caprolactam are effective and some commercial products of this type are now available. It is suggested that the pyrrolidone ring is the active site [(2)] enabling adsorption through the oxygen atom to the hydrate surface. It may also be possible for the pyrrolidone ring to become incorporated into the hydrate structure: either mechanism would lead to reduced crystal growth. Figure 2.29 shows how a kinetic inhibitor can depress the hydration.

Application of these polymers to a pipeline must be made at a location where the temperature is equal to or below the polymer cloud point temperature; otherwise, the polymer will precipitate at the point of injection and be ineffective. Kinetic inhibitors of a polymeric nature are not hazardous (the active ingredient is also used in personal care products).

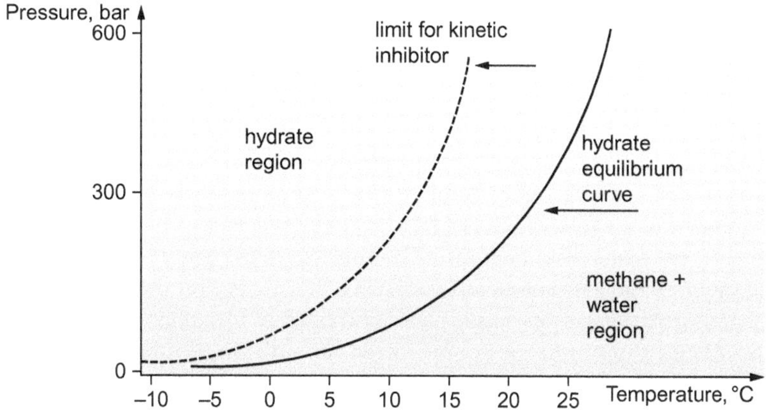

FIGURE 2.29 Pressure–Temperature Curve for Hydrate Formation in Methane–Water System Treated with Kinetic Inhibitor.

Anti-agglomerators (also known as crystal modifiers) prevent the agglomeration of small hydrate crystals, such that the solids that do form remain suspended and are transported through the system safely. Their performance is also characterised by a long induction period when there is little or no gas uptake. Addition rates are low (<1% based on water). Anti-agglomerators are surfactants or polymers that operate in hydrocarbon plus brine systems up to a water content of about 30%. A wide range of chemical types includes:

- Alkyl aryl sulphonates
- Alkyl polyglycosides
- Fatty acid alkanolamides
- Polyesters
- Alkylphenol ethoxylates
- Quaternary amines/quaternary phosphonium compounds

The quaternary amine products have received much interest and are in commercial use. One of the surprising benefits of these products is their ability to remain effective even during prolonged shut-in periods. They can be formulated with alcohols or mutual solvents and water. These inhibitors are typical emulsifiers and may function by partial formation of a water/oil emulsion at the point of addition.

Kinetic hydrate inhibitors and anti-agglomerators are usually formulated with other additives to enhance performance. Methanol and other low molecular weight alcohols are added as co-solvents and provide some of the benefits of thermodynamic inhibitors. In this case the active content of threshold inhibitor can be 10-20%. Although the costs of manufacture of threshold inhibitors are high per unit weight in comparison to methanol or ethylene glycol, the application dose is much lower and the overall economics now favour this new technology in many circumstances.

2.10.2.4 Application and Use of Hydrate Inhibitors

Hydrate deposits in pipelines can be extremely difficult to remove. Depressurisation at each end of a plug within a pipe may be needed in order to reduce the pressure to a value where the solid material will dissociate in the changed ambient conditions. For long gas or multiphase lines, continuous injection of methanol into the gas phase carries the solvent to wherever there is free water. Methanol partitions to the water phase and hydrate formation is stopped. The high volatility of methanol is considered an advantage, as it will be transported along the line. Methanol or ethylene glycol is sometimes deployed to wells that operate outside the hydrate region but that may develop hydrate conditions during shut-in periods. Chemical addition is then for a short period immediately prior to closure.

Methanol requirements to wet gas fields can be very significant. In the early 1990s one North Sea platform spent in excess of \$3 million on methanol and elsewhere the cost of methanol is above 5% of the total processing plant costs.[7] Field trials in the North Sea using threshold types of product have been successful and considerable savings are claimed compared to the use of methanol. In the high capacity wet gas line from the West Sole field, kinetic inhibitor at 3000 ppm was effective. The product was 15% active content and contained a corrosion inhibitor. Normal application rates for these threshold types are 1-2% based on water content. Many recent kinetic inhibitor trials have deployed a combination of thermodynamic inhibitor and kinetic inhibitors. For example, in one pipeline 1000 gallons/day methanol plus 1 drum of kinetic inhibitor have replaced the use of 2000 gallons/day of methanol. Frequently the two classes of inhibitor will be formulated together. Kinetic inhibitors are reported to perform increasingly well in high salt content water. This observation is the result of sodium chloride also migrating to the hydrate phase and affecting the morphology of hydrate formation.[7] A corresponding benefit is not achieved using methanol in high salt brine. Recent work suggests that multiphase flow will significantly impact the likelihood of hydrate formation. It is also recognised that nucleation of waxes or asphaltenes may precipitate hydrate formation and visa versa.

2.10.2.5 Hydrate Removal

Mono-ethylene glycol (MEG) and methanol are used to remove existing hydrate deposits. Treatments normally involve pumping a significant volume to the fouled area and allowing soaking for a few hours (caution: methanol is a precipitant of wax). Threshold inhibitors do not dissolve hydrates.

2.10.3 Paraffin Wax

One of the groups of products produced from the well are the higher molecular weight paraffins (n > 18). This fraction is stable and in solution in the reservoir. However, there are factors introduced in the production process that can change this equilibrium. The changes can be listed as:

- Changes in temperature as the oil is produced and passes through cold water or is cooled at the surface in long production lines
- Pressure drops allowing gas to come out of solution, which induces cooling
- Light ends removed as they flow better and are lost as the gas condensate fractions are removed
- Presence of fine-sized solids that can act as nucleation points for crystallisation
- High asphalt content contributing to wax formation as the particles can act as nucleation sites, although in other cases they can prevent crystallisation by interfering with crystallisation and flocculation processes

The most influential effect is the temperature drop and this, as shown in Figure 2.30, shows the potential problem areas of a production system where temperature losses are the greatest.

It must be kept in mind that production systems are very complex so this precipitation will occur over a range of temperatures that will be specific for a particular crude. The temperature at which this will occur is called the 'cloud point' or 'wax appearance point'. As the crystals start interacting with each other the viscosity will increase. Figure 2.29 gives a typical graph of viscosity as a function of temperature. The viscosity properties of crude behave in a Newtonian manner at higher temperatures. This means the viscosity is independent of the shear rate. However, as the first crystals start to appear and

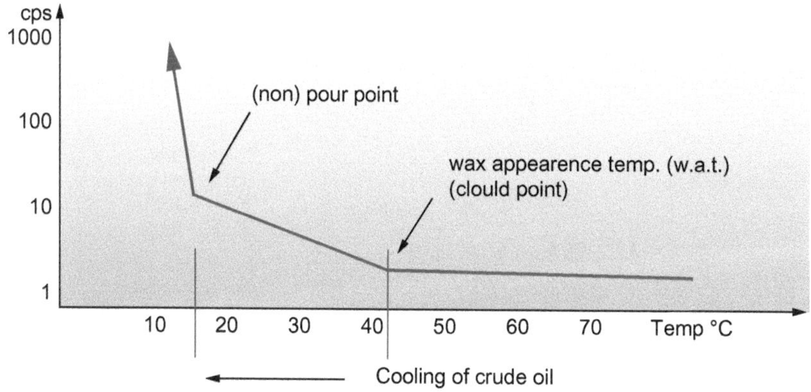

FIGURE 2.30 Graph of Viscosity of Crude as a Function of Temperature.

start interacting with each other the viscosity of the crude is increased. As the temperature (or pressure) is further reduced below the wax appearance point, viscosity continues to increase until, at the pour point, no further flow is possible. The stage at which the oil becomes very viscous is called the 'pour point'.

As the temperature continues to fall, the crystals, with included water and other organic compounds such as asphaltenes, will adhere to the cooler metal surfaces. Oil may also be entrained in the wax. It has been postulated that wax does not bond directly to metal but is actually held in place by the natural surface roughness.

There are two principal mechanisms by which wax deposition is thought to take place. Molecular diffusion transports wax, which is in solution, whereas shear dispersion transports precipitated, solid wax particles. This process of wax deposition is illustrated in Figure 2.31.

Wax deposits on the pipe surface increase the surface roughness of the pipe wall leading to an increase in frictional pressure drop and turbulent flow. However, the wax deposit acts as insulation of the pipe, keeping the temperature higher and thus reducing the wax deposition.

The problems of wax deposition are potentially more serious for the pipeline. The pressure drop required to move the oil may be greater than the available pressure from the pump system or the pressure drop may exceed the strength of the pipe. This particular problem, often associated with a shut down of the pipeline, may cause the pipeline to be blocked.

The production chemical treatment has to be designed to inhibit the wax crystallisation process and reduce the viscosity of the crude at a given

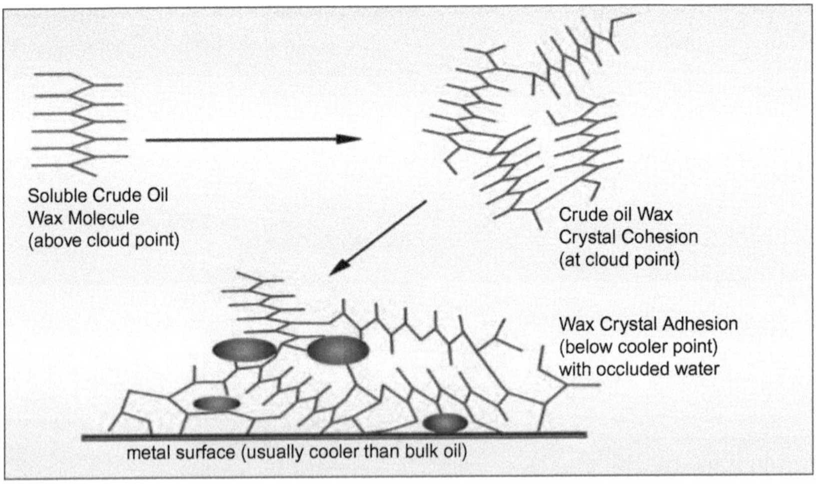

FIGURE 2.31 Wax Deposition Process.

temperature. The treatment can be designed from an analysis of the crude and a laboratory evaluation of the effect of potential treatments in the laboratory.

2.10.3.1 Characterisation of Wax

There is no standard definition of wax content, but it is generally accepted that straight and n-branched chain hydrocarbons, ranging from $C_{18}H_{37}$ upwards, are waxy. Crude oils vary significantly in hydrocarbon distribution, in 'wax content', and in propensity to deposit wax. Crude oils are in equilibrium within the reservoir but as the hydrocarbon and aqueous fluids are produced and transported this equilibrium is disturbed. The solubility of wax molecules in crude oil depends on the chemical composition of the crude, on the pressure, and the temperature. For a given crude, a drop in temperature or a drop in pressure can initiate the crystallisation process. Temperature is the most important factor. In the case of pressure change, it is the loss of light ends from the liquid phase that causes the change in solubility. Paraffin solubility is reduced by the presence of resins and asphaltenes but a significant reduction in wax deposition is observed for crudes containing a high proportion of asphaltic components. This is almost certainly due to the fact that asphalts exist as colloidal suspensions in crude oil and that the colloids interfere with the processes of paraffin crystallisation.

A number of models for describing wax formation have been published. Prediction of precipitation behaviour is difficult due to the large number of components (n-paraffins, iso-paraffins, naphthenes, and other aromatic and cyclic species) that can crystallize from crude oils. All wax models are based on empirical correlations and demand comprehensive analytical data for the aliphatic and aromatic components of crude to C50 max. Although these programmes give estimates of the amounts of deposited wax under specified conditions it is customary for these parameters to be also determined by experiment when specific field problems need to be solved.[7]

2.10.3.2 Laboratory Characterisation of a Crude Oil

It is important to obtain data on the rheological properties of the crude and the conditions under which wax will be likely to deposit. The following parameters are considered essential:

- Wax content
- Molecular weight distribution (distribution of n-paraffins and branched paraffins)
- Wax appearance temperature
- Asphaltene content

The above information will help define if there is going to be any problems and the likely demand for chemical additives. Crudes that are considered as problematic should be studied further and the following additional parameters determined:

- Pour point
- Wax deposition profile
- Pipeline restart properties (yield stress)
- Rotational viscometry

It is important to obtain fresh samples of crude. Ideally the sample should be maintained at the same temperature as it was when sampled. If it is allowed to cool and the wax crystals are allowed to form, re-heating will not necessarily reverse the crystallisation process. Also a problem with the sample is that as the pressure is lowered the volatile components are lost. The tests will not reproduce exactly what will happen in the field, but the results can be used to determine the extent of the potential problem and which treatment will be most effective.

Wax Content There are a number of chemical procedures available for determining 'wax' content. In one method solvent extraction is used to remove asphaltenes, other aromatics, and low carbon number alkanes. The wax fraction is then dissolved in warm chlorinated solvent and precipitated at very low temperature.

Molecular Weight Distribution Gas chromatography (GC) or gel permeation chromatography (GPC) is used to identify the paraffin molecular weight distribution in the crude oil. The proportions of n-alkanes, iso-alkanes, cycloalkanes, naphthalenes, asphalts, etc., can be estimated, although it becomes more difficult to differentiate the many isomers for carbon number C7 and above. The weight percent of C18+ fractions and therefore the wax content can be estimated from the chromatograms. A typical graph of distribution of different hydrocarbon chain lengths is given in Figure 2.32.

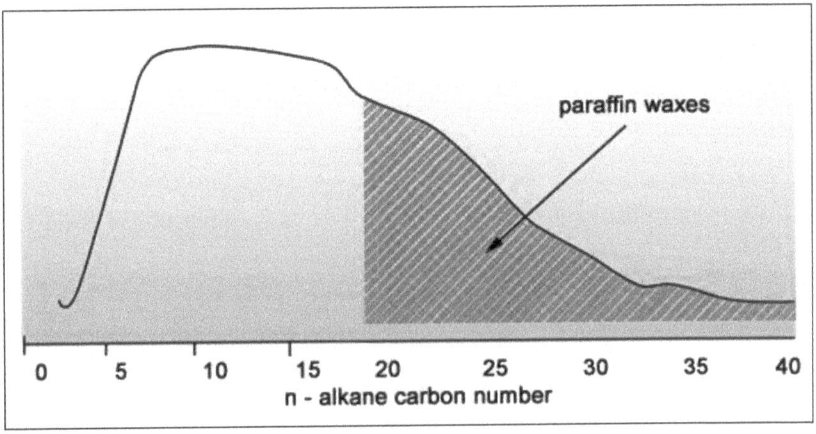

FIGURE 2.32 Typical n-Alkane Distribution of Dead Crude.

Wax Appearance Temperature (Cloud Point) This is arguably the single most important characteristic of a waxy crude oil, and it follows that it should be determined before conducting any wax deposition tests. The wax appearance temperature is the highest temperature at which wax crystals can be detected when an oil sample (previously heated to dissolve all wax solids) is cooled at a controlled rate. These crystals are very easily seen in refined, clear oil (hence cloud point) but indirect methods are used to find the value in a crude oil. The use of polarising microscopy, differential scanning calorimetry, filter plugging techniques, and Fourier transform infrared spectroscopy (FTIR) are all techniques used for wax appearance point. It is essential that the crude oil is heated to well above the estimated cloud point (e.g., 20 °F above) and held at elevated temperature for several hours before any test is commenced. Failure to do this will lead to incorrect readings and poor reproducibility.

Asphaltene Content The asphaltene content is defined by standard IP143 as that fraction of wax-free crude that is insoluble in n-heptane but soluble in hot benzene.

Rotational Viscometry A rotational viscometer measures the rheological parameters of a fluid. This would include the viscosity at a range of shear rates. These measurements can also be taken at different temperatures to determine the effects of temperature. Figure 2.32 shows a typical result for viscosity (determined at a fixed shear rate) as a function of temperature. It shows the viscosity increasing significantly as the wax crystals come out of solution at lower temperatures. The temperatures at which the viscosity character change relate directly to the wax appearance temperature (W.A.T.) and the pour point.

Analysis of the viscosity profile at a fixed temperature will show the oil behaving as a Newtonian fluid above the W.A.T. (viscosity the same over the shear rate range) and non-Newtonian below.

Pour Point For viscous crudes it is essential to understand the low temperature rheological properties. The pour point is one specific parameter that is usually measured and often forms part of quality specifications. Figure 2.33 illustrates the experimental set-up. The crude is initially pre-heated to 46 °C in a bath without stirring and conditioned for 30 minutes. IP 15 and ASTM D97 methods provide the lowest temperature expressed in multiples of 3 °C at which the oil is observed to flow when cooled under defined conditions.

Wax Deposition Tests Wax deposition rigs, particularly those capable of operation at elevated pressure, attempt to simulate the conditions experienced within the actual pipeline. The oil is conditioned above the wax crystallisation temperature. A 'cold finger' is inserted into the oil and the rate deposition of

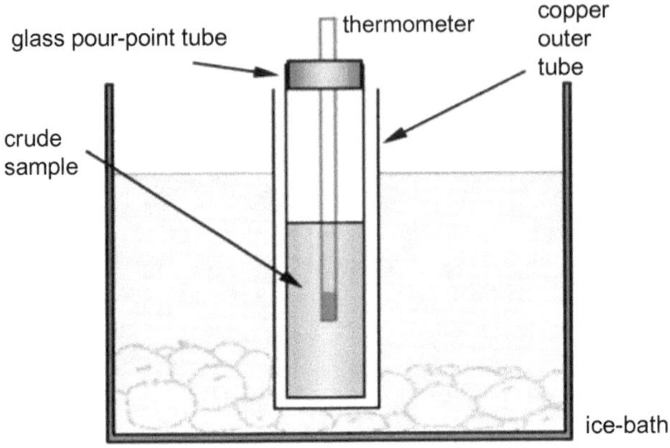

FIGURE 2.33 ASTM D-97 Pour-Point Apparatus.

wax on the clod finger is measured. A variant of this test is the capillary tube-blocking test.

Cold Filter Plugging Point This technique is widely used, especially on distillates. The temperature is measured at which a filter of standard mesh size is blocked by wax crystals from a crude or fuel. The cloud point can also be estimated from fuel samples.

Pipeline Restart Tests (Yield Stress) If restarting of a pipeline following a shutdown is predicted as problematic, yield stress test work should be undertaken. Laboratory pipeline restart test equipment is essentially a modified capillary tube-blocking rig with the capability to measure inlet and outlet differential pressure. The yield stress is measured as the point at which steady flow just begins after a low temperature shut-in period. Figure 2.34 gives a diagram of the equipment and a typical graph showing a build up of pressure to a point where the structure of the crystallised wax is broken down. This is referred to as the 'yield point' of the crude.

2.10.3.3 Wax Control Treatments

There are basically three options available for controlling waxy deposits: mechanical, thermal, and chemical. These can be further broken down to include:

- Mechanical
  - Pigging
  - Applied magnetic fields

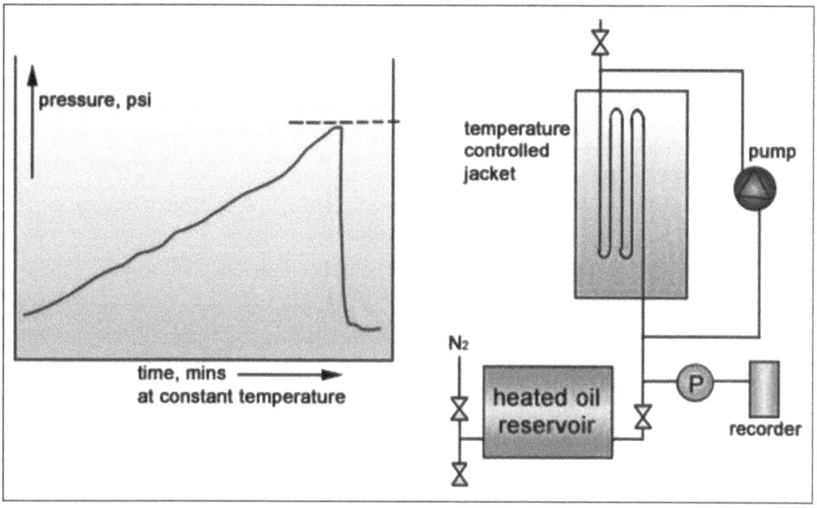

FIGURE 2.34 Laboratory Stimulation of Initiating Flow in a Pipeline.

- Thermal techniques
  - Insulation
  - Bottom-hole heaters
  - Hot water or brine
  - Hot oil
  - Heat of reaction between magnesium and hydrochloric acid
- Chemical
  - Wax solvent
  - Wax inhibitors (wax crystal modifiers)
  - Dispersants
  - Pour point depressants

Mechanical Methods Pigging is a very commonly used method. The pig is launched down the pipe and will mechanically scrape wax from the pipe wall and redistribute it within the crude in front of the pig. Wire line scrapers may also be employed. Such mechanical methods virtually guarantee that cleaning will be achieved but the processes can be expensive in time and labour costs. It requires specialist equipment, and there is the danger of the tools becoming stuck. A combination of pigging and other preventative measures may be optimal.

Thermal Techniques If the temperature of the oil can be increased, for example, by insulation of the pipes (useful for short sub-sea pipelines to a production facility) or by increasing flow, then wax layers will be removed as the surface deposits becoming softer and unstable. Increasing flow rate may, however, be accompanied by a significant increase in shear. Bottom-hole heaters are viable in certain situations. Hot oiling is the most commonly used technique for down-hole paraffin treatment in onshore fields. Hot crude is re-circulated to the zone that is affected by wax. Hot water or steam may also be used but there is the potential here to produce problematical emulsions. Periodic re-treatment is needed with hot-oiling and hot-water treatments. Another thermal approach is to use heat-liberating techniques such as magnesium bars followed by hydrochloric acid to produce a chemical reaction that liberates heat.

The use of these methods will assure paraffin removal. However, in order to prevent further deposition, the working temperature must be maintained above the melting point of the wax. Typically the process has to be repeated every 2-6 months. The heating process can create a fire hazard, which is a disadvantage, along with the costs and requirements for special equipment.

Chemical Solvents Dissolution of an existing deposit is viable providing the fouling is not too severe. A wide range of potential solvents is available as shown in Table 2.5. Hydrocarbon solvents that have historically been used include gasoline, condensate, toluene, xylene, and naphthas. Products with a high aromatic content will also act as a good solvent for asphaltenes. Aromatics and most alkyl-substituted aromatics are now classified as marine pollutants and must be labelled as a 'danger to the environment'. Phenyl-ester products and limonene have been examined as alternatives. Chlorinated solvents (trichloroethylene and other halogenated products) are not permitted for toxicological and environmental reasons. Similarly, carbon disulphide is not permitted due to high flammability and the off-gassing of toxic fumes.

Treatment will consist of filling the affected areas with the chosen solvent and static soaking for 6-24 hours. Long lines can be treated by slowly pumping the slug, but this approach can be logistically difficult and involves considerable downtime. A solvent is often included in the treatment of crystal modifiers.

2.10.3.4 Wax Inhibitors (Crystal Modifiers)

Chemical additives that are used to modify flow characteristics of crude oil can be classified in terms of the function they serve:

- Wax crystal modifiers (wax inhibitors)
- Pour-point depressants
- Dispersants

Chemically, polymer wax inhibitors and pour-point depressants are similar and the functions of wax inhibition and pour-point depression can usually both be achieved with the same product. Many workers describe all these 'inhibitors' as flow improvers. Dispersants act in a different manner. Many commercial wax inhibitors also contain dispersants.

TABLE 2.5 Solvents for Wax Treatments

Compound	Flash Point, °C	Comments
Hot Crude	Variable	Low cost but can present fire hazard
Gas Oil	Variable	Safer than crude and widely used
Alkyl Aromatics	Variable but can be >30	Toxic to marine life
Naphtha	Variable	Toxic to marine life
Limonine	48	Extracted from pine trees. Toxic to marine life
Phenyl ester		Environmentally better but expensive
Xylene	27	Potential fire hazard but most commonly used
Toluene	6	Toxic and low flash point
Trichlorethylene		Toxic to aquatic life
Carbon disulphide (CS_2)	-30	Not used due to high flammability and formation of toxic fumes

Wax Crystal Modifiers These inhibit deposition by co-crystallising with the wax crystals and so prevent the formation of structured lattices at the pipe wall. This is illustrated in Figure 2.35. The process is thought to be one of habit modification by inclusion of regular crystallisable structural sequences in the crystal lattice, where further growth is prevented by the disruptive effect of relatively bulky and non-crystalline side groups. The modified wax agglomerations are less likely to form a massive cohesive deposit and will not readily adhere to surfaces. It is essential that the additive be added to the oil at a temperature at which both the additive and the wax are completely in solution, i.e., in an amorphous state. Two key qualities are required for a successful crystal modifier:

- It should precipitate in crude oil at about the same temperature as the wax, and
- It should contain structural sequences capable of co-crystallisation with wax.

Successful crystal modifiers are polymeric or high molecular weight organic molecules with both straight chain hydrocarbon and polar groups within the same molecule. Key characteristics of these polymers are the presence and frequency of pendant side chains along the backbone and the molecular

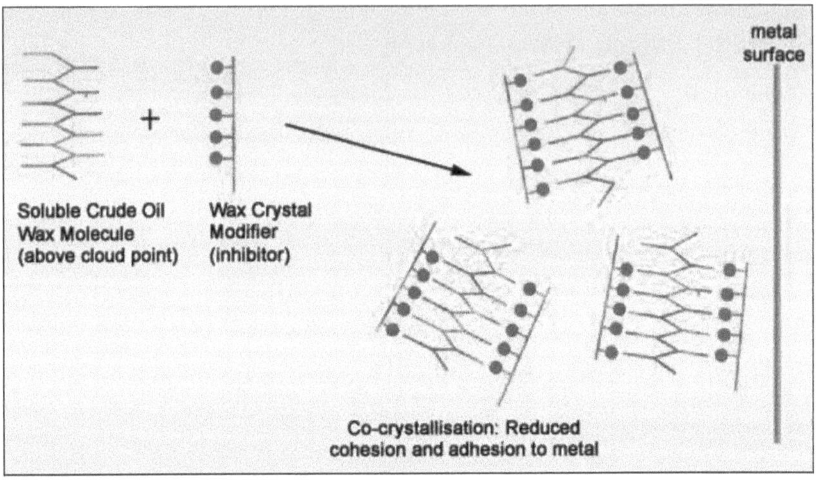

FIGURE 2.35 Mechanism of Wax Crystal Modification.

weight. It is the interaction of polar groups on the side chains with the crude paraffin molecules that largely determines if the product will be effective.

A typical polymer that is widely used is EVA co-polymer (ethylene–vinyl acetate) with the formula – $(\text{-}CH_2\text{-}CH_2\text{-})_m$– $(CH_2\text{-}CH(OCOCH_3)\text{-})_n$-. The molecular weight is in the range of 10,000 to 30,000 so the number of repeating units is around 100 to 300. The vinyl-acetate content is between 18-40%. The acetate side chain gives it a branched character. Other polymers widely used are alkyl-acrylate polymers, maleic and succinic acid esters, and co-polymers.

The polymers are normally applied as a 5% solution in an aromatic solvent to aid rapid dispersion into the oil phase. The product is applied on a continuous basis to achieve maximum benefit and should be added to the warm/hot crude at a temperature above the wax appearance point. Typically the addition points are at the wellhead and prior to entry into the pipeline to the refinery. Surfactants are commonly added as co-additives (see also dispersants). Application levels between 100-1000 ppm of formulated product are typical.

Pour-Point Depressants Increases in viscosity coupled with crystallisation of waxes at low temperature hampers the transportation of crude oils through long distance pipelines. Additives that depress the pour point and improve the flow characteristics at low temperatures are described as pour-point depressants. It is likely that pour-point depressants also act as crystal modifiers but the converse is not necessarily true. These additives are polymers with n-alkyl chains (primary or side groups) and other polar groups. The suitability of the molecule for a given application will depend upon a number of features including:

- Molecular weight
- Length of the alkyl chains
- Distance between the pendant side chains
- Distribution of monomers in a co-polymer
- Amorphous or crystalline nature of the polymer, etc.

The length of the alkyl chain is important as it determines the temperature at which these chains will crystallise and adsorb on to the surface of the wax particles in the crude. The molecular weight affects the solubility in the crude so it would be expected that similar products of differing molecular weight would not give the same performance in any given oil.

Ethylene-vinyl acetate polymers are widely used as pour-point depressants. These co-polymers are available in a wide range of physical and chemical properties. Both the number of side branches and the length of the branches are important. Acrylates and maleic/succinic/fumaric acid esters and co-polymers are also recommended.

The treatment tends to be specific for a particular crude and laboratory evaluations are critical in determining the best treatment. Figure 2.36 shows the effect of treatment on the viscosity/temperature curve for two inhibitors. In this example Product B is preferred.

The pour point of a New Zealand crude was lowered from 32 °C to 11 °C by the use of a 720 ppm (summer) to 1280 ppm (winter) pour-point depressant. Field application rates were found to be lower than laboratory simulations (7).

Dispersants Adsorbed films on a metal surface reduce the adhesion of paraffins to that surface(7) and that the nature of the compounds adsorbed determine wettability characteristics. Wax dispersants probably function by

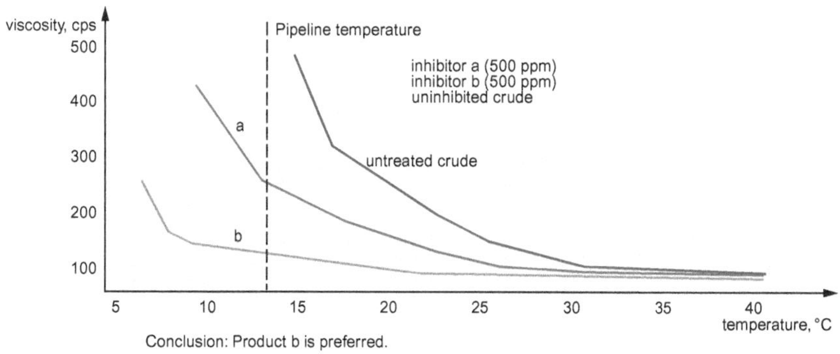

FIGURE 2.36 Viscosity–Temperature Curves to Determine Optimum Treatments of Pour-Point Depressants.

adsorbing and water wetting the surface of the pipe and by adsorbing to pre-existing wax crystals, thereby reducing the tendency for them to stick together. The overall effect is less accumulation on the sidewall. A good dispersant formulation will also function to penetrate accumulated deposits of wax, adsorbing on individual particles, and enabling them to move freely into the surrounding oil. Wax dispersants may be dosed continuously for inhibitory effect or by batch treatment to achieve remedial benefits.

Anionic surfactants, including olefin sulphonates, polyolesters, and other powerful wetting agents, are typical additional ingredients of wax dispersants. Formulations may also contain primary wax inhibitors (10-30% total active concentration).

2.10.4 Alsphatene

Asphalt is part of the heavy and polar fractions of crude oils. Asphalt has two components:

- Asphaltenes comprise condensed aromatic and naphthenic molecules of molecular weights ranging from several hundred to several thousand and are characterised by a significant heteroatom content: nitrogen, sulphur, and oxygen. Figure 2.37 gives the complex structure.
- Resins are similar to asphaltenes but have alkyl, non-polar side chains. The two components form a stable association where the resin acts as a dispersant, generating colloidal charged particles that repel each other. Figure 2.38 illustrates how the two components interact to form a stable structure.

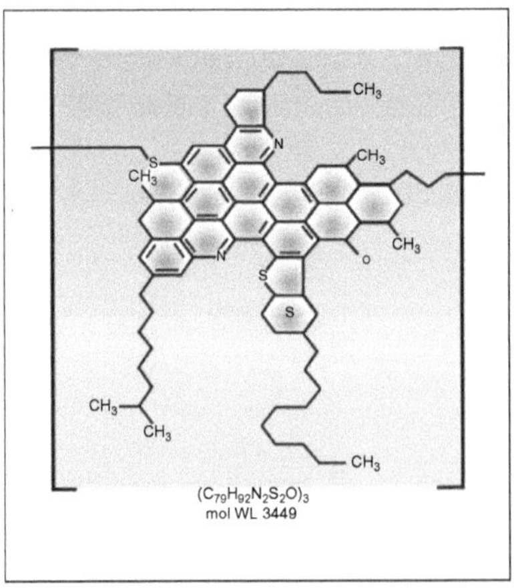

FIGURE 2.37 Asphaltene Structure.

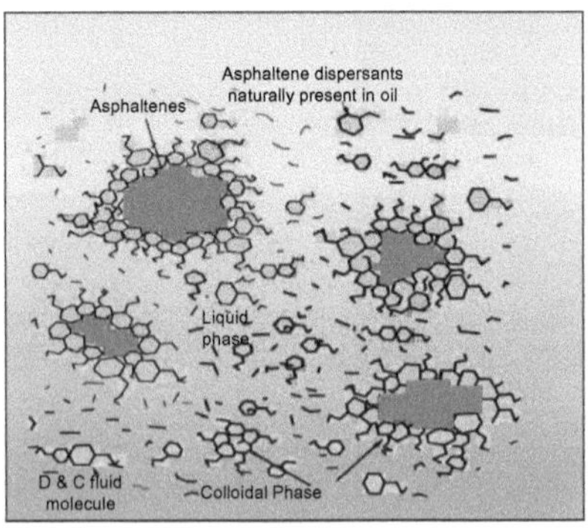

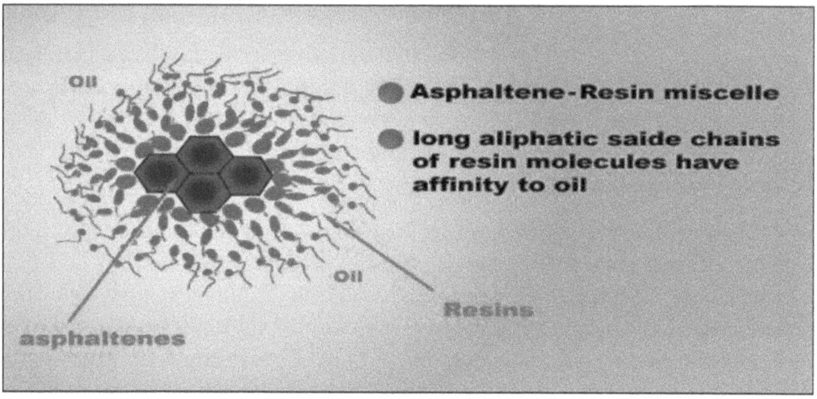

FIGURE 2.38 Asphaltene–Resin Association.

They can be isolated from crude oil using selective solvent extraction techniques. The asphaltenes are classified as the n-heptane (or n-pentane) insoluble fraction. Resins can be described as the fraction that is soluble in n-heptane, toluene, and benzene but insoluble in ethyl acetate at room temperature.

The ratio of the asphaltenes to the resins is an important feature that gives the system stability and if it is less than one (more resin than asphaltene) the system tends to be stable. In production systems, changes in temperature, pressure, and chemical composition of the crude, combined with streaming-potential effects in the well tubing, affect asphaltene stability. Deo [(1)] discusses the complexity of the precipitation processes.

Pressure and temperature have an influence on the stability of asphaltene. A number of workers have reported that in oil well tubulars deposition occurs below the depth at which the bubble point for the crude appears (the pressure at which a gas phase separates). This observation is attributed to the different compressibility of the lighter ends and the heavier components of the crude oil. As the crude approaches its bubble point, the relative volume fraction of the lighter ends within the crude increases. Its effect is similar to adding a light hydrocarbon to a crude, resulting in asphaltene de-peptisation. Above the bubble point (depth) the low molecular-weight alkanes are released from the liquid into the gas phase, which reduces the tendency for asphaltenes to be further de-peptised in the crude.

Acids, particularly in the presence of iron and CO_2, can also contribute to the precipitation of asphaltenes.

The operating problems caused by asphalts are similar to waxes but are in fact more difficult to predict and to treat.

2.10.4.1 Solutions to Asphaltene Deposition Problems

There are no satisfactory and reliable predictive tools available that will define the severity of asphaltene deposition. Inhibition is unreliable, and most of the literature describes remedial treatments (rather than preventative measures) at specific field locations. Techniques that have been recommended for asphaltene control include:

- Adopt preventative action through the addition of continuously dosed inhibitors, dispersants, or solvents to reduce the deposition rate
- Chemical cleaning of wells and surface equipment through the re-circulation of asphaltene dissolvers (solvents) through the affected area
- Avoid the mixing of crude streams; the blending of crude feedstocks is a common cause for asphaltene precipitation; light, non-asphaltic crude is a possible precipitant for heavier crudes
- Temperature and pressure manipulation of the produced fluids in order to minimise the occurrence of conditions that have been determined to promote asphaltene deposition and thus extend the on-stream efficiency of well production and equipment
- Mechanical cleaning of the wells and surface equipment; this includes the use of wire line methods and the opening up of vessels, e.g., separators, and literally digging out the accumulated material

2.10.4.1.1 Asphaltene Dissolvers

The discussion concerning the selection of a solvent is comparable to the one for wax solution. Pyridine and carbon disulphide and halogenated hydrocarbons are efficient but are toxic, and this has prevented their use in the field. Simple aromatics such as toluene and xylene have been shown to be generally effective and are relatively inexpensive but have handling disadvantages (flash points 5, 27 °C, respectively). Higher alkyl-substituted benzenes have higher flash points but lower solvency than xylene. Most substituted aromatic solvents are now classed as marine pollutants. Bi-cyclic and polycyclic solvents are claimed to be effective. Their structures are given in Figure 2.39. Many of the higher molecular weight solvents have flash points above 61 °C and are therefore not regarded as highly flammable.

Pyridine and dimethyl pyridine have been generally found to be best solvents for asphaltic deposit.[7] Apart from the toxicity disadvantages it was found that the pyridines also attacked elastomeric O-rings that swelled badly. The methyl naphthalenes caused practically no damage to the elastomer.

To remove asphaltic deposits in flow lines, solvents are either batch treated or re-circulated to the affected zone for 8-24 hours. For treatment of tubulars, the product is used neat or diluted with crude and re-circulated via the annulus or bull-headed.

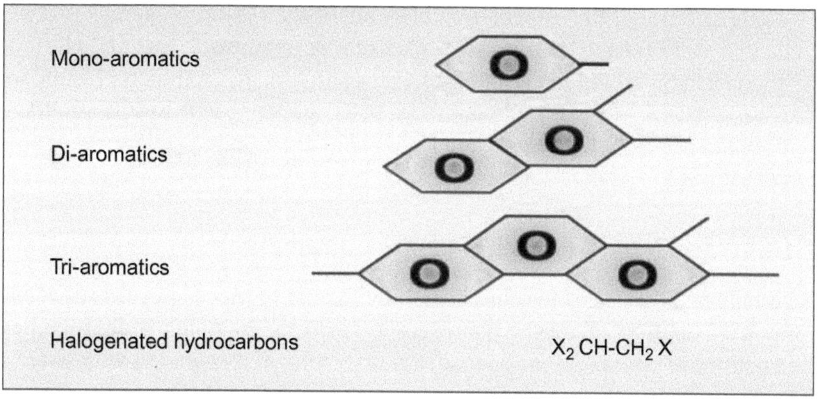

FIGURE 2.39 Asphaltene Solvents.

2.10.4.1.2 Asphaltene Inhibitors/Dispersants

The design objectives of an 'inhibitor' should be to stabilise the colloidal association and reduce the rate of particle flocculation, leading to a reduction in deposition to surfaces. Examples include alkyl part esters of phosphoric acid:

$$\mathrm{R\,Alkyl - O - P(O) - (OH)_2}$$

Alkyl-aryl sulphonic esters (e.g., didecylbenzene sulphonic acid, DDBSA) are common dispersants:

$$\mathrm{X - SO_3 - C_6H_6 - C_{12}Alkyl}$$

Continuous circulation of hot (deasphalted) oils or aromatic solvents can reduce or eliminate asphaltic problems by diluting the crude.

2.10.4.1.3 Combined Wax Asphaltine Emulsion Treatments

Chemical vendors offer products that are combined wax/asphaltene inhibitors. They are deployed in the same way as asphaltene dispersants are used and are useful if an operator has both paraffin and asphaltic problems. Emulsions are stabilised by asphalts stiff emulsion character). A chemical vendor should be able to develop a combination product for the removal and prevention of stiff emulsions but optimisation may require significant on-site testing.

2.10.4.1.4 Field Application of Asphaltine Solvent

The success of any field treatment depends on a number of factors, but it is crucial that the position of the deposit is known with some accuracy and that a reasonable estimate of the quantity of asphaltic deposition can be made. This is certainly important for treating a fouled tubular (an example is given below).

Compatibility of the solvent with elastomers and plastics materials that are contained within the system must be checked. Aromatic solvents and particularly condensed aromatics are aggressive against many common elastomers that are used in oilfield application as seals, gaskets, etc. Swelling and/or embrittlement of elastomers are possible even after relatively short contact times.

2.10.5 Emulsion Management

An emulsion consists of two incompatible phases; in this case, oil and water, co-existing as droplets of one phase in the other phase, which is described as the continuous phase. Oil produced from reservoirs in association with natural formation water or mixed formation water/injection water will, due to mechanical agitation, form an emulsion naturally, stabilised by emulsifiers occurring naturally in the produced fluids. The stability of this emulsion is strongly dependant on the detailed chemistry of the water and oil and will vary in character as the field matures with a consequential change in the ratio of produced oil and water. In addition to emulsion, there will be some free water, the proportion of which will usually increase as the water content increases.

Failure to separate (or dehydrate) the oil/water mixture efficiently can result in a number of problems, including:

- Overloading of surface separation equipment
- Increased cost of pumping crude that contains significant water
- Significant flow line or tubing pressure resulting from high viscosity emulsions
- Increased vessel heating costs
- Risk of corrosion in export lines, including sub-sea pipelines and at refineries
- Stock tank bottoms form thick sludge, difficult to remove
- Refineries can only accept oil of low basic sediment, water, and salt content; typical maximum acceptable salt content at a refinery is 25 pounds of salt per thousand barrels of crude
- Risk of catalyst poisoning at refinery

The significance of this aspect of production is reflected in the very large chemical market for demulsifiers, desalters, and oily water (reverse demulsifiers) chemicals that account for approximately 4% (in value) of the world oilfield production chemicals market. This section will describe how chemicals are used to resolve these problems.

2.10.5.1 Emulsion Stability

Emulsion droplets can be of all sizes, from fairly large (visible) to sub-micron in size. It is found that for primary produced oil, the continuous phase is the oil, and the dispersed phase is the water. This is an example of 'normal' water in oil (w/o) emulsion and may contain from trace to about 90% water. An emulsion is thermodynamically unstable but the presence of naturally occurring emulsifiers and other solids can significantly increase the stability. The emulsion is unstable when the droplets can fuse to form bigger and bigger droplets. Some emulsions of produced oil and water may be tight (difficult to break) or loose (easy to break).

Oil in water (o/w, or the opposite, water in oil) emulsions are typical of water effluent streams where the oil content is low and which must be further treated to reduce oil content before discharge of water to the environment. Low salt content primary crudes are often found to be oil in water emulsions.

It is not always obvious from field samples if an emulsion is a w/o or an o/w type. This can be established by adding either water or kerosene to the sample: whichever fluid dilutes and mixes with the emulsion is the continuous phase. Adding water to oil continuous emulsion will not result in dilution and the water will simply drop to the bottom as a free, un-combined phase. Milk is a water continuous emulsion and is easily diluted with water.

Crude-oil emulsions are formed due to the mixing of water and oil, vigorous agitation, and the presence of emulsifying agents in the crude. The amount of mixing that occurs in the reservoir is relatively minor and even in the production tubular string there may not be adequate turbulence to cause the formation of a stable emulsion. However, as the crude passes through chokes, wellhead valves etc., extreme mixing conditions are experienced at the pressure gradient and this forms new w/o interfaces. This emulsion formation is illustrated in Figure 2.40 where the different types of emulsions are also illustrated.

Mechanical agitation is the basic cause of emulsion formation but various components of the crude then influence the stability of the emulsion:

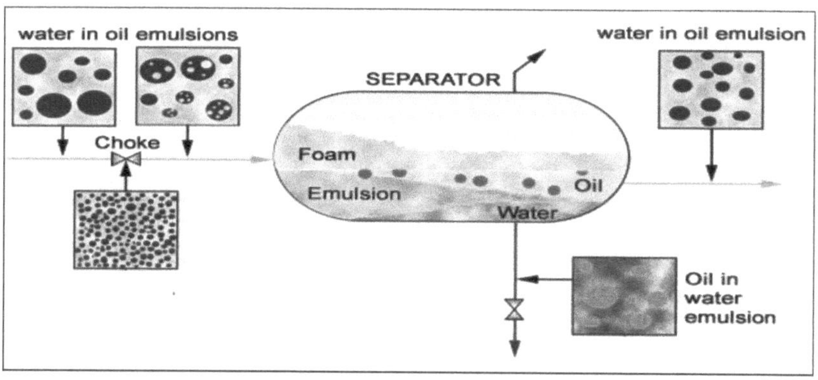

FIGURE 2.40 Diagram Illustrating the Different Sites of Emulsion Formation and Nature of Emulsion.

- Natural surfactants: these can be organic acids or esters, heterocyclic nitrogen compounds, or various oxidised hydrocarbon species and act to reduce surface tension slightly (in conventional surfactant terms they are only weakly adsorbed)
- Resins are more interfacially active and absorb at the newly created interfaces
- Asphaltenes accumulate at the interface
- Paraffin wax crystals also precipitate during production
- Inorganic fines (silt, clays, scale) will be present and can stabilize interfacial films; oil wettable solids such as iron-sulphides and oxides form stable emulsions; these films can therefore have a strong mechanical strength
- Introduced production chemicals such as corrosion inhibitors, biocides, scale inhibitors, or wax inhibitors can be described as surfactants in chemical character so may also act as emulsifiers

Soluble components of the associated water phase can also affect emulsion stability. High levels of soluble calcium or magnesium can increase stability. Fluids produced at a lower than normal temperature (seasonal effects, heater failure) will be more viscous and the resulting emulsion will be less easy to break. Production through long sub-sea lines results in cooler fluids but in this case the extended time the fluids are contained within the pipeline can de-stabilise the emulsions. Steric repulsion of adsorbed surfactants, interface rigidity, and solids stabilisation are properties that can explain emulsion stability.

2.10.5.2 Emulsion Instability and Demulsification

Crude-oil emulsions are not thermodynamically strong and given adequate time will separate into oil and brine phases. This is in contrast to some well-known domestic examples of emulsions, e.g., mayonnaise that will remain stable for years.

The de-stabilisation of water in oil (w/o) emulsions occurs in stages:

- Flocculation stage
- Creaming or dropping of water stage
- Coalescence or breaking of free water stage

These stages are illustrated in Figure 2.41.

The flocculation stage involves the association of droplets, including the smallest sub-micron size drops, into clusters but without coalescence or settling. This is followed by ‘creaming’, which is the term often used to describe the rise or fall of the dispersed phase droplet clusters under the influence of gravity. There must be a difference in density for creaming to occur so in a crude-oil system it will be the dense brine phase droplets that will drop toward the bottom of the vessel. The result of this process is a concentration of dispersed phase drops in an emulsion that is particularly rich in the dispersed phase. At this point there is considerable risk of inversion of the emulsion, e.g., from 90% w/o to 10% o/w.

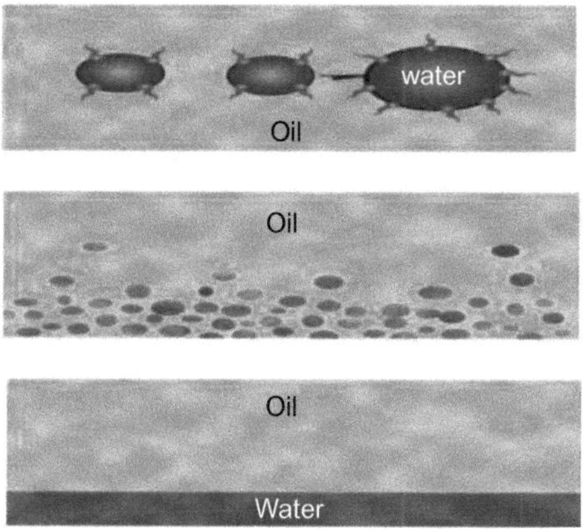

FIGURE 2.41 Demulsification Stages of Water in Oil Emulsion.

The final stage of emulsion breaking occurs when the droplets coalesce. Drainage of the continuous phase leads to rupture of the interfacial films between drops and an overall reduction in free energy (lower surface area). Enhancement of the destabilisation processes can be achieved in a number of ways:

- Increased temperature
- Centrifugation
- Electrical methods
- Chemical treatment
- Increased resonance time

By increasing the temperature, the dispersed phase droplets that are in continuous Brownian motion are encouraged to strike each other more often and with greater force. There is therefore an increased likelihood of coalescence. This process is called perikinetic flocculation. Heat also reduces the viscosity of the continuous phase giving higher film drainage rates. Hydrocarbons expand with a temperature rise at a greater rate than water, so heating the emulsion also increases the differential density therefore encouraging settling. Heat also increases the solubility of the natural emulsifiers and emulsion stabilisers in the liquid phases, which results in the removal of these materials from the surfaces and destabilisation of the emulsion.

Centrifugation is used to artificially increase the G-force and increase the concentration of the dispersed phase in the 'creamed' emulsion layer. Breaking of the interfacial films can also occur by this process.

2.10.5.3 *Electrical Demulsification*

High-voltage electrostatic separators are widely used to treat water in oil emulsions in which water (the dispersed phase) has a much higher conductivity or polarity than oil. The applied field causes a dipole to form on water

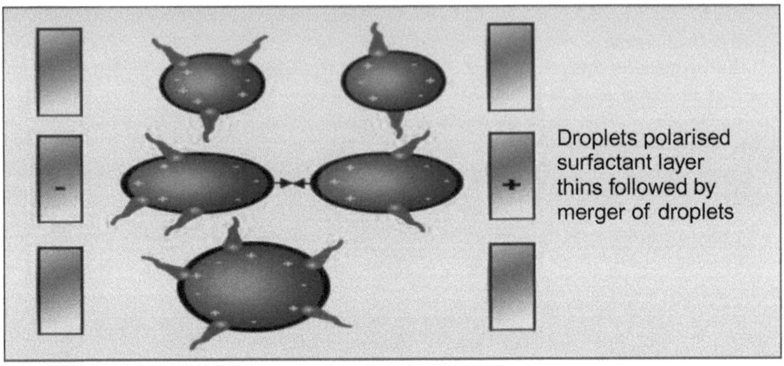

FIGURE 2.42 Destabilisation of Emulsions by Electrostatic Field.

drops that are further de-stabilised by becoming elongated. Interaction between droplets creates forces of attraction, which lead to aggregation and coalescence. This mechanism is illustrated in Figure 2.42. The electric fields used to resolve water in oil (crude) emulsions are high voltage alternating current (AC) electric fields. By contrast, oil in water emulsions respond to direct current (DC) current, leading to enhanced flocculation of droplets through electrophoresis.

Emulsions are thermodynamically unstable, and given sufficient time will be resolved. The resonance time between wellhead and final separation varies considerably from site to site and affects the selection of demulsifier.

2.10.5.4 Chemical Demulsifiers

Chemical demulsifier formulations are used throughout the world to improve emulsion-breaking processes. Jjoblom [(1)] discusses mechanisms for demulsifier products. Present-day demulsifiers are blends of highly sophisticated organic compounds with surface-active characteristics. Demulsifier formulations are prepared from non-ionic, cationic, or anionic surfactants. The action of a demulsifier is to destabilise the emulsion.

Figure 2.43 illustrates how the solubility of the emulsifier (hydrophilic-lipophilic balance, HLB, value) influences whether the emulsion is water in oil or oil in water. This figure shows how a polymer of intermediate HLB value can displace the emulsifiers and therefore thin the emulsifier concentration and cause coalescence.

Figure 2.44 also shows how the HLB value influences both the stability and the type of emulsion as well as where the polymeric demulsifiers are in relation to the other emulsifiers.

2.10.5.5 Demulsifier Selection Procedure

Selection of the optimum chemical normally is made by ‘bottle tests’ and experienced technicians conduct a series of optimisation experiments using

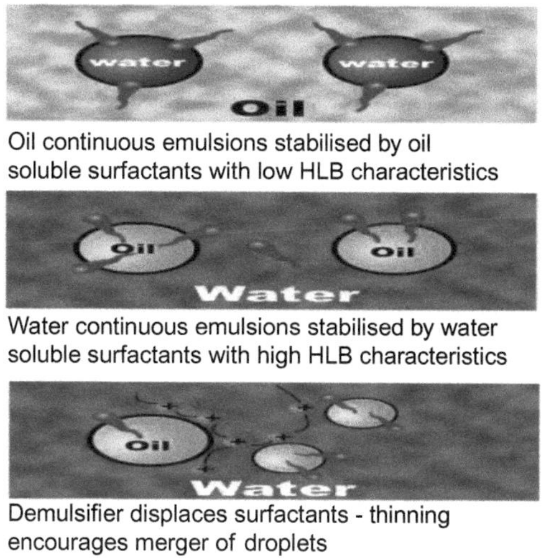

FIGURE 2.43 Diagram Illustrating the Different Sites of Emulsion Formation and Nature of Emulsion.

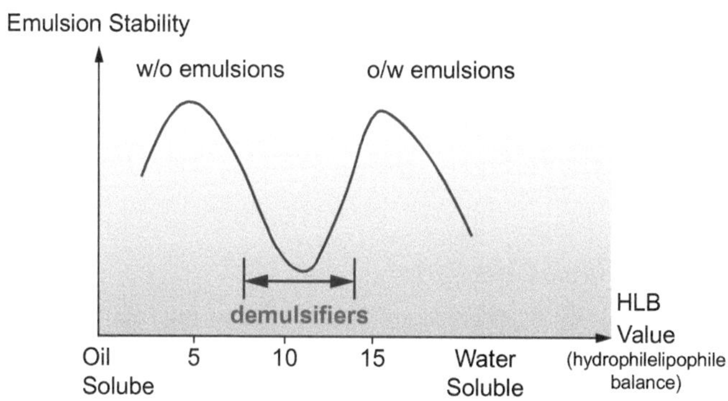

FIGURE 2.44 Relationship Between Emulsion Stability and HLB Value of Surfactants.

the large number of surfactant bases available. It is most important that testing is conducted on representative fresh samples of emulsion, as aged (oxidised) oils will not give the same result.

A good sample of fresh emulsion is collected from the produced fluid stream and is then divided into several test bottles (sometimes pre-calibrated). A wide selection of demulsifier bases, alone and in combination, is tested. The preferred product will have been chosen because of many factors such as the brightness of the oil, clarity of the water, interface quality between the oil and water, required operating temperature, settling time, salt content, the BS&W (bottom sediments and water) specification of the export crude, and cost. Laboratory results on fresh crude are a good indication of field performance.

Oil in water emulsions may be tested by substantially the same procedure, using water to dilute the emulsifier bases. Temperatures are generally lower and successful chemical concentrations would normally be less.[7,8]

2.10.5.6 Dehydration Plant Systems

Oil and water produced by the well move from the wellhead and separator through the treating facilities, and finally the crude is transferred to stock tanks or pumped via the custody meter and pipeline. The final crude should ideally contain no solids or water or emulsion.

2.10.5.7 Separators

These vessels are usually the first piece of equipment in a process system and are designed to remove the gas from the oil and the water from the oil. Initially the gas must be removed, which is done in a two-phase separator. The purpose of the three-stage separator and free-water knockout vessels is to remove any remaining free water, thus preventing overloading of the dehydration equipment. This is particularly important where the crude requires heating, as it takes nearly four times more heat energy to achieve the same temperature in water as in oil. It is common for knockout drums to contain filter material to help remove entrained oil from the water.

The wet oil and emulsion then feed into the dehydration system proper, which can be either a wash tank (gun barrel) or a heater treater, or a chem-electric treater.

2.10.5.8 Wash Tank-Gun Barrel

This vessel is basically a settling tank fitted with an internal or external boot. The tank is known as a gun barrel and the boot as a wash tank. Although these units are not common, they are found in older locations, and are an excellent example of how emulsions can be treated. The barrel or inlet pipe acts as a gas separator and anti-surge device, thus reducing turbulence in the main tank. A spreader plate allows the slowly moving emulsion to be distributed evenly into the water phase allowing it to be washed and encouraging the water particles to increase in size. The settling time is important here as water droplets that increase in size can fall by gravity allowing clean oil to rise and be removed for storage and shipping.

2.10.5.9 Heater Treaters

Heater treaters come in many sizes and shapes, from as small as 50 b/d to as large as 10,000 b/d throughput. Of all the different types of dehydration equipment available, heater treaters are probably the most versatile. They are normally delivered packages complete with thermostats, level controls, and burners simply requiring connection to the system. Vertical treaters are more common for smaller operations. For larger operations a horizontal treater is preferred. The vertical treater operates first by pre-heating the emulsion in the external heat exchanger before the fluid flows down to the spreader below the firebox. As the fluid is heated around the firebox coalescence of the water takes place in the filter tray section, free water being drained from the base and the hot oil exiting through the heat exchanger as it is discharged to storage.

2.10.5.10 Chemelectric Treaters

These treaters are now very widely used. Like the heater treater, they are fully automatic, available in various sizes, and normally arranged horizontally. In the first stage of operation, the principle is the same as the heater treater: gas leaves the heater end, avoiding breakout later in the process. The hot wet oil enters the electrical coalescing section via an emulsion spreader to give uniform flow. A high-voltage alternating electrical field excites the particles, thus bringing them together and accelerating coalescence of the small droplets of water. The placement of electrodes and selection of voltage is adjusted according to the requirements. Chemelectric treaters are available without heating sections if the inlet temperature of the oil emulsion is sufficient for electrostatic dehydration.

2.10.5.11 Pipelines and Storage Tanks

In some oilfield systems where long inter-field, trunk lines, and large storage tanks are present, demulsifiers are injected at the wellheads and use the residence or transit time in the pipeline for gravity separation of the oil and water. As fluids arrive at the storage tank the separated water is drained from the tank and the 'dry oil' is then exported. Free water in long pipelines is a potential cause of bottom-of-line corrosion.

2.10.5.12 Desalting

Crude oils contain traces of impurities such as iron-sulphide, sand, silt, and water. Entrained water normally contains salt in the form of chloride, which when heated to high temperatures (average distillation temperature 650 °F) can hydrolyse, releasing hydrochloric acid. This acid along with others present in the crude (napthenic) can cause serious corrosion problems within the refinery pipework.

Parameters such as crude temperature, composition of the wash water, the degree of mixing, and system settling time are variables that determine the quantity and type of demulsifier (desalter) used.

2.10.5.13 Effluent Water Treatment (Crude-oil Production)

Residual oil in water is separated in an effluent treatment plant. Effluent from onshore or offshore production units or refineries must be environmentally acceptable for disposal or be suitable for re-injection into the producing reservoir or aquifer. Any effluent that is discharged must conform to a current maximum specification for oil in water of 10 ppm in accordance with OSPAR (Oslo-Paris) regulation. Care is required when interpreting o/w data based on infra-red adsorption, as dissolved organic acids (not removable) are included, giving artificially high 'oil' contents.

2.10.5.14 Stability of O/W Emulsions

Oil droplets and solid contaminants generally have a residual surface negative change at neutral pH in oilfield water-treating systems. This net negative surface charge is caused by increased hydration of (most) cations compared to anions; the latter are more polarised and are preferentially absorbed at the surface. Surrounding this surface negative charge is a layer rich in counter-positive ions – an electric double layer. Long-range repulsive forces due to the electric double layers are believed to be the main stabilising influence in o/w emulsions.

2.10.4.15 Destabilisation of O/W Emulsions – Flocculation of Oil Droplets

The use of hydrocyclones or centrifuges is increasing. Their use is becoming more popular, since they require little or no chemical addition. Gas flotation of oil from water is still widely practised and does require the use of chemical flocculants. These are sometimes known as reverse demulsifiers or oily water clarifiers. Chemical-assisted flocculation of oil droplets is achieved by charge neutralisation using synthetic water dispersible polymers of high functionality. Cationic polyelectrolytes such as polyamines are able to overcome the negative repulsive forces at the oil drop surface.

2.10.4.16 Filter Aids: Polyelectrolytes and Inorganic Salts as Flocculants/Coagulants

Polymeric flocculant molecules of molecular weights from 104-107 are used to induce aggregation of suspended particles and oil droplets and hence increase capture efficiencies. These polymers are typically used at the level of 0.1-10 ppm and vary widely in chemical type charge type and charge density. Iron III or aluminium III compounds are capable of forming large flocs of hydroxide salts in solution, which are themselves capable of entraining or flocculating indigenous particles.

Aluminium compounds include:

- Aluminium Sulphate $Al_2(SO_4)_3.18\ H_2O$
- Sodium Aluminate $Na_2Al_2O_4$

Aluminium salts are not normally used for seawater as they only form hydroxide flocs in the pH range of 6-7, whereas seawater has a pH of about 8.

Iron compounds include:

- Ferric Sulphate Fe $(SO_4)_3.3H_2O$
- Ferric Chloride $FeCl_3.6H_2O$

Iron compounds form a hydrated ferric hydroxide floc suitable for sea in the pH range 8-11. It should be noted that ferric-chloride solutions are highly corrosive. In practice it is very often essential to use a combination of polymers and inorganic coagulants to achieve optimum efficiency in the system. Other chemicals (e.g., biocides, scale inhibitors, corrosion inhibitors, etc.) used to treat water for the purpose of injection to a subterranean reservoir will be described in the section on water injection.

2.10.5.17 Effluent Water Treatment Plant

It is most important that the effluent water leaving the treatment plant conforms to the local environmental regulations and guidelines. An API separators unit is the most commonly found item in all areas of the world, and has been modified locally to improve its efficiency. The unit relies on gravity separation: oil floats to the surface and sludge falls to the base, while passing through bays normally made from concrete. The residence time needed for separation dictates the size and number of bays. Oily water enters by an adjustable inlet and the floated oil concentrate is either removed by moving skimmer pipes or by a static adjustable (height) pipe at the outlet bay. The clean water is taken off at mid-level. In some units, sludge collectors are fitted.

2.10.5.18 Plate Separators

These units are available in many designs (atmospheric or elevated pressure), and are based on the API separator principle. To speed up the separation process tilted plates are located within the vessel at an angle of 45° to the horizontal. The operation of these units depends on the density differences between oil, water, and sludge. Oil rises and coalesces on the underside of the tilted plate where it is collected. Sludge collects at the base, while the clean water passes between the plates. These units are more efficient and use less space than the basic API separators, but the tilted plates are prone to corrosion.

2.10.5.19 Dissolved Air Flotation (DAF)

In this type of unit, saturating a pressurized partial stream of water with air/gas and then releasing this fluid into the raw water inlet of the flotation vessel at atmospheric pressure generates very fine bubbles of air or gas. The pressurised recycled water generally varies between 10-30% of the total stream, and on release to atmospheric pressure produces bubbles in the range of 40-60 micron. With the help of polyelectrolytes, the bubbles then attach to the oil floc, which rises rapidly to the surface and is skimmed to waste. Selection of polyelectrolyte and optimisation of concentration are important to maintain high efficiency.

2.10.5.20 Induced Air Floatation (IAF)

This unit introduces air or gas into the oily water stream that then forms small bubbles to aid froth generation. Figure 2.45 illustrates such a unit. These units are less expensive and occupy less deck space than the DAF type, but are less efficient. Both types of unit described are usually arranged in series with up to five cells to clarify the wastewater. As water passes from one cell to another its contaminant loading is reduced. Both depend on careful level control and efficient skimming. Neither DAF nor IAF units are entirely suitable for floating production platforms, and can be inefficient during bad weather. On fixed offshore platforms the swaying action of the structure can upset liquid levels.

2.10.5.21 Filters

The types of filters used are very similar to those used for the solids removal in water injection systems, but in this application they are also capable of residual oil removal along with solids. The degree of contamination and throughput determines the type of filter needed: pressurised mixed media units being most common. The filter units are described in more detail in the water injection section.

2.10.5.22 Coalescers (Figure 2.46)

These are mainly based on cartridge systems, and the media used for coalescing is typically 'hyperfine glass fibre' (1-10 micron in diameter) or synthetic polymer material. It is essential that efficient prefiltration of the oily water

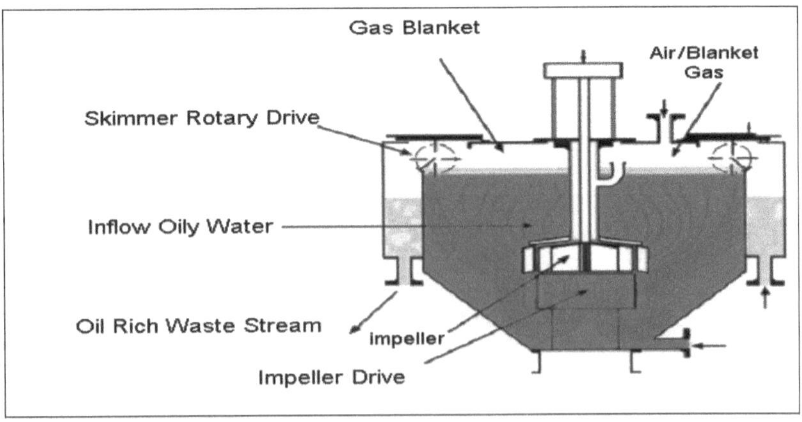

FIGURE 2.45 Diagram of Induced Gas Flotation Unit.

FIGURE 2.46 Spooled Mare's Tail *(Source: Opus [8]).*

stream is undertaken since the coalescers act as very good filters and can rapidly blind with solids.

2.10.5.23 Activated Carbon Adsorption

Although activated carbon is a relatively expensive, the adsorption process is efficient at removing pollutants, especially organic and toxic compounds. Very low contaminant levels are achieved (micro grams per litre). Pre-treatment with filtration and flotation is required to prevent unwanted pollutants.

2.10.5.24 Hydrocyclones

These units have become very popular in the North Sea due to their very low space weight ratio. The process utilises gravity separation: the effluent water enters the cyclone and is induced to spin, creating high 'G' forces on the water forcing the oil to the central core. The units are able to reduce the oil loadings to below the 40 ppm current discharge content, but are sensitive to fluctuations in flow rate. Oil droplet size must be greater than 10 microns for separation.

2.10.5.25 Centrifuges

Centrifuges have been used for many years for the separation of heavy oils and water (tar sand oil), but until recently have not been supplied to the North Sea fields on a large scale. They operate by utilising centrifugal forces, which separate both oil and solids from the water stream. Oil droplets as small as 1 micron can be separated.

2.10.5.26 Biological Units

Aerobic biological treatment is a process used to remove organic pollutants from wastewaters so as to reduce the BOD (biological oxygen demand) down to a level that fulfills local environmental legislation. The design criteria for these units are dependent on various factors including temperature, pH, and the presence of toxic compounds. Bacteria are allowed to grow either on large sand granules or plastic cups situated in towers or tanks onto which the contaminated water is sprayed. Bacteria degrade organic contaminants. Flow rate and residence time of the unit controls the final effluent quality.

2.10.6 Corrosion Management

Corrosion can be defined as the destructive attack of a metal by chemical or by electrochemical reaction with its environment. Within the petroleum producing industry electrochemical corrosion is responsible for millions of US dollars of costs associated with the breakdown of equipment caused by the loss of metal from inner and outer surfaces. Corrosion affects all stages in the exploitation of crude oil, ranging from drilling of the wells through the crude production separation process, export pipelines, storage tanks, refineries, and finally, even the internal combustion engine.

Electrochemical corrosion occurs at the solid/fluid interface in water, water/oil, and gas systems. It can occur in H_2S (sour) systems, in CO_2 (sweet) systems, or in a combination of both. The consequences of electrochemical corrosion can be severe and include, general metal wastage (general corrosion), embrittlement of steel, and surface cracking. If left untreated, any of the above processes can lead to equipment failure.

All forms of corrosion are found in this industry, which means that engineers have to be ever-vigilant in preventing equipment downtime and selecting the most cost-effective solution, whether this be optimum metal selection, optimum process design, or optimum chemical inhibition. This section will explore these options.

2.10.6.1 Electrochemical Corrosion

For an electrochemical reaction to occur, a number of requirements must be met. There must be:

- Surface anode
- Surface cathode
- Electrolyte present
- Some form of external connection between the cathode and anode

The anode is the area on the steel surface where oxidation processes occur. In the case of a steel coupon immersed in brine, the basic reaction at the anode is the conversion of iron metal into ferrous ions:

$$Fe^{o} \rightarrow Fe^{2+} + 2e - \text{(oxidation)}$$

There is therefore removal of steel at the anode. At the cathode reduction reactions occur. The electrolyte provides the source for hydrogen ions that are converted to atomic hydrogen. This has the property of being able to migrate through steel and as it recombines with other hydrogen atoms to form hydrogen gas causes problems such as blistering and embrittlement:

$$H^+ + e^- \rightarrow H$$

$$2H_2O + O_2 + 4e^- \rightarrow 4OH^- \text{(reduction)}$$

The electrolyte transports corrosive materials (such as dissolved gases) to the surfaces and acts as a medium for the transport of ions. The electrolyte carries ferrous ions from the anode and hydroxyl ions from the cathode: once the solubility product is exceeded these ions combine to form solid ferrous hydroxide, which concentrates at the boundary of the anodic and cathodic regions:

$$2Fe^{2+} + 4OH^- \rightarrow 2Fe(OH)_2 \text{(corrosion product reaction)}$$

Any connection between the anode and the cathode, which is not through the electrolyte, is an external connection and is normally the body of the steel. The value of the potential between the anode and the cathode is proportional to the severity of corrosion. Pure water is a poor electrolyte and is in fact only mildly corrosive to steel. In this case, iron reacts to form ferrous hydroxide and hydrogen. This causes a slight increase in pH and as soon as the solubility product is exceeded a layer (passive) of ferrous hydroxide is deposited on the reactive surface, thus preventing further corrosion. Oilfield brines are significantly more corrosive than freshwater. This is because many of the dissolved components of these brines enhance general corrosion.

The rate of general corrosion of ferrous metals is strongly influenced by a wide number of factors:

- Acidity: The aqueous acidity is influenced by the content of organic acids (typically 100-1000 ppm) and by dissolved carbon dioxide. Corrosion is increased by acidity. The presence of carbonate minerals in the formation acts as a buffer and reduces corrosivity.
- Temperature: At low pressures corrosivity increases with temperature up to about 65 °C and then decreases somewhat due to deposition of corrosion product films. At higher pressures, corrosivity continues to increase with temperature up to much higher values (200 °C).
- Pressure: Increasing pressure results in an increase in stress related failures.
- Fluid velocity: High fluid velocities remove protective films (corrosion product films, scale deposits, or inhibitor films).
- Chloride content: Pitting and other localised corrosion is enhanced by high chloride content (high salinity) assuming that some oxygen is present. The effect is particularly noticed at high temperatures.
- Imposed electric currents: This can be a problem if electric heating is used. Direct current and low frequency alternating current are likely to increase the general corrosion rate.

- Influence of dissolved oxygen: It is difficult to control corrosion in the presence of oxygen, and even low levels (>5 ppb) can be quite corrosive. Water can contain up to 10-ppm oxygen under normal conditions of T, P, and salinity. Oxygen reacts at the cathode (depolarisation) and causes oxidation of ferrous iron to ferric hydroxide (rust):

$$O_2 + 2H_2O + 4e^- \rightarrow 4OH^-$$

$$4\,Fe(OH)_2 + O_2 + 2H_2O \rightarrow 4\,Fe(OH)_3$$

- Oxygen can only be effectively removed in a closed system when the above reactions will continue until all oxygen is consumed. Oxygen exclusion is important for the successful use of other corrosion control methods. Vacuum deaeration and scavengers (bisulphite) are used to reduce the oxygen to acceptable levels in injection water systems (below 5 ppb).
- Influence of dissolved carbon dioxide: Carbon dioxide/bicarbonate ions are present in all oilfield waters and even at low partial pressures can result in attack on low alloy steels. This is referred to as 'sweet' corrosion. The depolarising reaction of carbon dioxide is:

$$2CO_2 + 2H_2O + 2e^- \rightarrow 2HCO_3{}^- + H_2$$

- Flow rate is an important factor in CO_2 water corrosion as at high flows it is possible to remove protective carbonate films. Stainless steel is resistant to CO_2 under low-pressure conditions. Knowledge of system parameters allows prediction of corrosion severity and models such as those developed by De Waard and Milliams are helpful.[7]
- Influence of hydrogen-sulphide: H_2S is relatively soluble in water creating a weak acid that is corrosive. In the absence of oxygen, H_2S will attack low-alloy steels. In the presence of carbon dioxide and oxygen, corrosion is enhanced and even acid-resistant alloys will be attacked. Blistering and sulphide stress cracking (SSC) are common and serious corrosion forms.

The National Association of Corrosion Engineers (NACE) standard MR 0175-88 indicates that a gas phase partial pressure of H_2S equal to 0.05 psia is sufficient to lead to SSC in low-alloy steels. At low temperatures (<65 °F) and in steels not heat treated SSC is less problematic. Hydrogen penetration reduces metal ductility and creates high internal pressures.

2.10.6.2 Corrosion in Oilfield Systems

General corrosion or uniform corrosion is the most 'acceptable' form of corrosion that can usually be predicted and controlled by the use of chemicals or optimum materials selection. When general corrosion occurs the anodic and cathodic areas are continually changing, which causes all areas to be evenly corroded. General corrosion can be evaluated and monitored in the laboratory and in the field. Other forms of corrosion are difficult to predict and difficult to detect in the field.

Localised corrosion is more serious and can lead to early failure. All or most of the metal loss occurs at discrete positions. Pitting is a form of localised corrosion and may occur as a consequence of a non-homogeneous surface becoming exposed under layers of foreign matter or at breaks in surface coatings. Most of the factors that apply to general corrosion also apply to localised corrosion.

Crevice corrosion is a form of pitting, which is induced usually at a contact between nuts, bolts, or rivets and the normal metal surface, and is the result of local and minor differences in the environment (local corrosion cells established). SSC caused by sulphides can occur in acidic waters (CO_2). Dissolved sulphide causes a corrosion cell to develop that produces iron-sulphide (FeS) and hydrogen. Penetration of the metal by hydrogen reduces metal ductility and creates high internal pressures at metal imperfections. Heat-treated high-strength or low-alloy steels are susceptible. Hydrogen embrittlement, in which the metal becomes glass-like (brittle), is a consequence of hydrogen invasion. Optical and scanning electron microscopy are valuable tools for examining cracking phenomena.

Galvanic corrosion is the result of immersing dissimilar metals in electrolytes. At the junction of the two metals there will be a flow of electrons, which leads to electrochemical pitting. Nickel and other corrosion-resistant alloys can increase the likelihood of galvanic corrosion or embrittlement of steel. Erosion corrosion is attack accelerated by high fluid velocity that can either physically remove protective films or mechanically disturb the surface itself. Particles of sand or iron-sulphide cause erosion corrosion even at moderate velocities. Cavitation is a special form of attack caused by the collapse of bubbles formed at areas of low pressure in a flowing stream.

2.10.6.3 Corrosion Control – Non-Chemical Methods

A wide range of control measures is available but not all of them can be adopted in any given environment. These measures include:

- Engineering solutions
- Coatings
- Cathodic protection
- Treatment of the water

Metals are available for most corrosive environments in oilfield systems but the usual limitation is their cost. High-chrome steels, such as 'duplex and super duplex', resist acid gases and are frequently used in HT-HP setups. Corrosion-resistant metals are now more commonly used for pumps, meters, and auxiliary fittings. Titanium and Monel are resistant to saltwater attack and are used in water injection systems.

Painting and galvanising are examples of coatings that have long been applied to reduce corrosion. The corrosive environment is therefore separated from the metal at risk. Inorganic coatings such as vitreous enamel or glass are very resistant but brittle, and cement is popular for water lines. Organic coatings include the use of paint, lacquer, and plastics and are frequently used to protect tubing and tanks. With the exception of sacrificial coatings, any imperfection or break in the surface finish (holiday), which exposes base metal, will result in serious local corrosion possibly in the form of severe pitting. Coatings are liable to physical damage through running wire line tools through the tubing.

Cathodic protection is the application of an electric current to the surface of the metal at risk such that the whole surface becomes cathodic. The current is applied externally to a special anode, which ensures that the original steel surface becomes entirely cathodic, including areas previously anodic. A 'protective current' can also be obtained from a sacrificial anode that is coupled to the structure to be protected. The anode is selected to be more base in the galvanic series, e.g., zinc rather than the steel. Cathodic protection is used to protect the submerged areas of water tanks, ships, pipelines, and offshore installation structures.

2.10.6.4 Corrosion Control – Chemical inhibitors

If it is possible to remove the main corrosive agents from the water by physical processes, and then downstream corrosion will be reduced. A corrosion inhibitor is a substance that, when added to an environment, decreases the rate of attack by the environment on a metal. Inhibitors can be inorganic, organic, polymeric, simple, or complex formulations. Passivating inhibitors such as chromates, molybdates, silicates, and phosphates form a strong non-corrosive oxide film with steels that causes the surface to be cathodic. They are not used to protect internal surfaces in the oil production industry as corrosion control by passivating inhibitors is only possible if the surface is 100% covered. For the purpose of this discussion, two classes of inhibitors will be discussed: oxygen and sulphide scavengers and organic film forming.

Oxygen Scavengers Corrosion of steel in brine is enhanced by dissolved oxygen. Neutral water of low salt content in equilibrium with air at 70 °F (21 °C) will contain about 9 ppm of dissolved oxygen. The concentration of oxygen decreases with increasing salt concentration and increasing temperature. The most commonly used scavengers of oxygen are sodium and ammonium bisulphite solutions. They react to form stable sulphates:

$$O_2 + 2Na_2SO_3 \rightarrow Na_2SO_4$$

The reaction rate of sodium sulphite with oxygen at low temperature is also very slow, so a catalyst is usually added. The treatment rate is 8-ppm sulphite per 1-ppm oxygen. While cobalt, manganese, and copper salts are good catalysts, cobalt at 0.04% gives the greatest increase in reaction rate. Copper should not be added to steel systems because it lowers the hydrogen over-voltage. Ammonium bisulphite is supplied as a 65% solution and requires no catalyst. Sixty ppm of ammonium bisulphite as supplied will treat 9-ppm oxygen. Oxidising agents interfere with these reactions.

Sulphide Scavengers Producing wells are sometimes characterised by having high and potentially hazardous concentrations of sulphide species. The sulphide ion is a natural constituent of many produced fluids. The activity of sulphate-reducing bacteria, particularly in water-flood operations, also produces sulphide. It is not unusual for sweet production systems to turn sour with time as a consequence of increased sulphide concentrations. The use of the word 'sulphide' implies all three water-soluble forms:

- H_2S (in low to neutral pH)
- HS^- (neutral to alkaline pH)
- S^{2-} (alkaline pH)

Curves for distribution of ionic species as a function of pH are given in Figure 2.47. Sulphides are unwanted constituents of crude oil and water. Their presence in the separated crude has a detrimental effect on the sales value and performance of refineries. In acid form, hydrogen-sulphide is toxic to health and accelerates corrosion processes. In production systems with natural sour production processing, equipment is specified to resist the corrosive nature of sulphides.

An ideal chemical scavenger would:

- Provide complete, efficient, and reliable scavenging of harmful sulphide species
- Produce reaction products that are inert

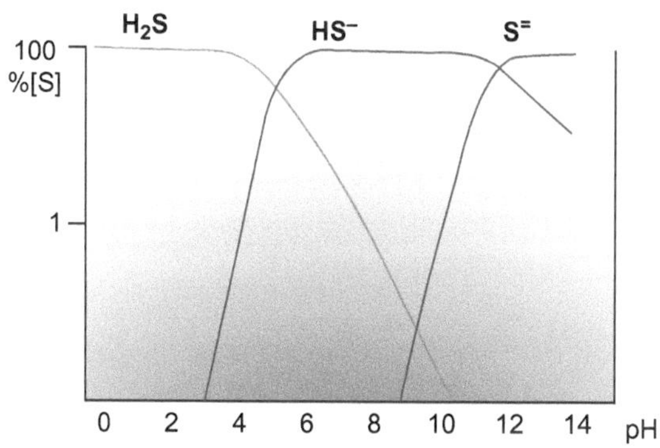

FIGURE 2.47 Distribution Curve of Hydrogen-sulphide Ions as a Function of pH.

- Be efficient under all likely conditions of temperature, pressure pH, and in the presence of other chemical additives
- Be non-corrosive (reaction products should also be non-corrosive)
- Be non-hazardous to the health and safety of personnel and would not pollute the environment
- Be readily available and economical to use

Unfortunately, there is no individual sulphide scavenger that meets all these desirable characteristics and a compromise is required.

Oxidants The basic reaction of oxidising scavengers can be summarised as:

$$O_2 + 2H_2S \rightarrow 2S^o + 2H_2O$$

Most strong oxidants are technically effective. They are rapid acting, water-soluble, and irreversible in action. All are potentially highly corrosive to ferrous metallurgy. Hydrogen peroxide and t-butyl peroxide have been used. Chlorine dioxide, ClO_2, is a strong oxidant that has the advantages of also being a biocide and does not form harmful by-products. Aqueous sodium chlorite at $pH > 9$ is used as the precursor. The reaction with H_2S is rapid and irreversible:

$$5H_2S + 8ClO_2 + 4H_2O \rightarrow 5SO_4^{=} + 9Cl^- + 18H^+$$

Iron Chelate A chelated iron product has been used as a regeneratable oxidant. Precipitation of FeS or $Fe(OH)_3$ is avoided due to the chelant:

$$Fe^{3+}EDTA + H_2S \rightarrow S + Fe^{2+}EDTA$$

The system is effective over a wide range of pH values.

Amine Types Simple amines react with sulphides/H_2S to form acid/base salts in a fully reversible reaction. Ethanolamines have been used extensively to remove H_2S from gas streams in sweetening towers.

Aldehydes H_2S reacts with aldehydes across the $C=O$ double bond in a reversible process. The most common use is of formaldehyde. Glyoxal, a dialdehyde, is claimed to have superior scavenging properties to formaldehyde. Gluteraldehyde and acrolein are sometimes used but all aldehydes are restricted by toxicity concerns and by temperature, ionic strength, and pH limitations. Acrolein is a powerful lachrymator.

Solid Precipitation Scavengers These are used in H_2S scrubbing towers. Iron sponge (solid) is probably the earliest commercial sulphide precipitating agent. The material is formulated by impregnating wood chips with iron-oxide and is primarily used in drilling fluids:

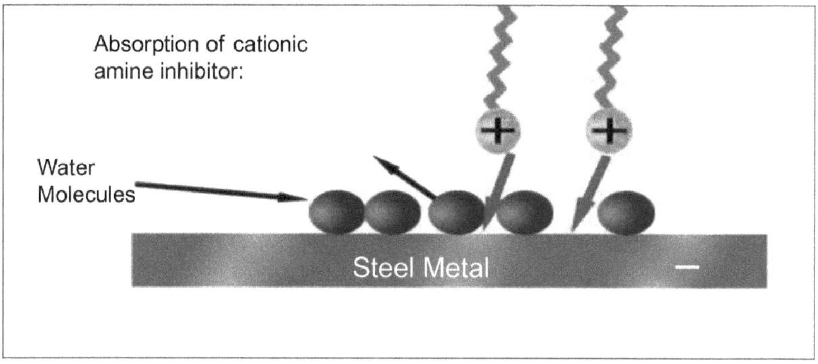

FIGURE 2.48 Action of "Filming" Inhibitors.

$$Fe(Oxides) + H_2S \rightarrow FeS + H_2O$$

Zinc compounds, as slurry in contactor towers, rapidly precipitate as sulphide when H_2S is present in the gas stream. There is an environmental impact problem in disposing of the waste zinc sulphide:

$$ZnO + H_2S \rightarrow ZnS + H_2O$$

Organic Inhibitors This class of inhibitors function by adsorbing on the metal surface, therefore, displacing water and forming a waterproof layer. The efficient use of organic inhibitors depends on prior removal of oxygen from the system through the use of oxygen scavengers.

Cationic Nitrogenous Inhibitors These are the main active constituents in most oilfield inhibitor compositions. The terminal amine group is protonated or carries a positive charge so is cationic at use concentrations. The cationic polar part of the molecule is attracted and chemisorbed to the negatively charged sites on the steel surface, and the fatty (hydrocarbon) part of the molecule is exposed to the crude oil and acts as a barrier to water. The inhibitor will displace any pre-existing water molecules on the steel surface because they have lower interaction energy. This action is illustrated in Figure 2.48.

Examples of cationic nitrogenous inhibitors include:

- Fatty amines and quaternaries (R = long chain hydrocarbon)

$R\text{-}NH_2$	$R_2.NH$	$R_3.N$	$(R_4N)^+$
Primary	Secondary	Tertiary	Quaternary

- Diamines (primary and secondary amine functions)
- Amides/Imidazolines
- Ethoxylated amines and imidazolines

- Salts of nitrogenous bases with carboxylic acids
- Heterocyclics

Diamines such as n-stearyl-1,3-diamino propane and its salts are more popular than primary amine salts as components of corrosion inhibitors.

Imidazolines are formed by the condensation of polyamines, usually DETA, with fatty carboxylic acids. Amides are formed from the same ingredients but under differing reaction conditions. Fatty carboxylic acids that are used include low-cost tall oil (C18+minor rosins) and oleic (C18 unsaturated). Amides and imidazolines are widely used in oil-dispersible inhibitor formulations. To prepare water dispersible versions, imidazolines, and fatty amines are reacted with acetic or propionic acid. Imidazolines and primary amines generally become more efficient as the chain length is increased.

Salts, formed by the neutralisation of an (fatty) amine group with fatty acids (tall oil, naphthenic, dimer, and trimer acids), are widely used in solvent-based formulations and form highly persistent films to the metal. Persistency is a particularly useful property if the inhibitor is not added on a continuous basis (e.g., slug-dosed). Organic inhibitors form films that are water repellent. Certain hydrocarbon components of the crude will also co-adsorb with the inhibitor to improve the barrier effect.

Non-nitrogenous Organic Inhibitors Certain alkyl half esters of phosphoric acid are considered [(3)] good inhibitors of carbon dioxide and are used in oil production and condensate streams. Napthenic acids have been used for crude-oil systems because they have lower toxicity characteristics but the adsorbed inhibitor film is more easily removed by water.

2.10.6.5 Formulation and Properties of Organic Corrosion Inhibitors

Corrosion inhibitor formulations are usually mobile liquids. They are designed to be dosed on a continuous basis to flowlines or by batch treatment and can be water or solvent based. The main components of water-based corrosion inhibitors are typically:

- Primary inhibitor bases 10-30%
- Surfactant 0-2%
- Demulsifier 0-2%
- Inorganic synergist 0-6%

2.10.6.6 Application of Corrosion Inhibitors

Continuous Down-hole Treatment Corrosion inhibitors can be pumped directly to the bottom of the oil or gas well via a macaroni or kill string where well completions allow. If a bottom-hole injection valve is fitted, then a solution of corrosion inhibitor is placed in the annulus and a make-up inhibitor solution added with a suitable pump. Continuous treatment through an injection line is the preferred method of treatment to give reliable corrosion control. Typical applications are:

- For oil wells: 25-1000 ppm based on water production
- For gas wells: 25-2000 ppm based on water production

For this downhole treatment application, the inhibitor should have good long-term stability as a solution in diesel (for example).

Batch Treatment Batch treating of oil or gas wells involves the use of a calculated volume of a corrosion inhibitor, either neat or as a solution/dispersion in solvent/water, plus a flushing liquid introduced into either the annulus or the tubing. In an oil well without a packer, the inhibited fluid can be recirculated from the annulus through the tubing and back to the annulus to ensure good film thickness. In a gas well the inhibited fluid is pumped into the tubing and allowed to flow under gravity to the bottom by shutting in the well for up to 24 hours. Wells with a standing column of fluid may require a 'weighted' inhibitor formulation. Typical inhibitor applications are:

- For oil wells: 500-50000 ppm, twice per week to once per month
- For gas wells: 5-50%, diluted, weekly to every 2-months

Squeeze Treatment Squeeze treatment consists of injecting corrosion inhibitor into the tubing with a high-pressure pump, followed by several tubing volumes of overflush, such that the inhibitor solution penetrates the formation face. After adsorbing to the surfaces of solids in the formation, the inhibitor is slowly produced back into the wellbore with the well fluids. Treatment lifetimes are much longer than with batch treatments. With squeeze operations, it is particularly important that the inhibitor does not 'gunk' in the solvent as this would result in formation damage. There is also a potential risk of emulsion block formation. Squeezing of corrosion inhibitor can be carried out as a 10% solution in diesel (for example) with a diesel overflush. A typical treatment would require 100 barrels of corrosion inhibitor with an overflush of 500 barrels. Core compatibility must be established prior to treatment design and application.[7]

Full-Tubing Displacement Similar to a squeeze, this method is used when there is a risk of formation damage. It involves the injection of a high concentration slug of inhibitor (e.g., 10%) in hydrocarbon solvent or crude to the production string. Enough volume is used to fill the tubing and the well is shut in for a limited period, e.g., 1 hour, before being put back in production.

2.10.6.7 Application of Corrosion Inhibitors to Flowlines and Surface Equipment

For surface equipment, it is operationally best to apply corrosion inhibitors continuously as far upstream of the problem area as possible. Wellhead addition is commonplace in many locations. Initially it may be desirable to dose the inhibitor at a concentration in excess of that normally required so as to establish good film coverage. Thereafter, a rate of 2-100 ppm is necessary, according to the corrosivity of the fluids. Injection to oil export lines is critical. The injection equipment and metering pumps provide the correct amount of chemical to the problem area. Again, it is essential that the chemical be dispersed as quickly as possible toward the metal.

Water-Injection Systems Water-injection systems, whether they are for water disposal or for waterflood operations, frequently experience corrosive attack from the water. This attack can sometimes be controlled by continuously treating with a corrosion inhibitor. Water-soluble inhibitors have been used for this application. A review of corrosion control in injection wells is provided by Patton [(6)].

Figure 2.49 summarises the points where corrosion treatment are applied in a typical production system.[7]

2.10.6.8 Corrosion Monitoring

The objective of monitoring is to obtain information about the corrosion process and thereby to assess options for controlling corrosion and its consequences. The main reasons for monitoring include security of assets/investments (economic reason) and security of personnel safety (health and safety reason). Corrosion control means controlling the rate of corrosion to

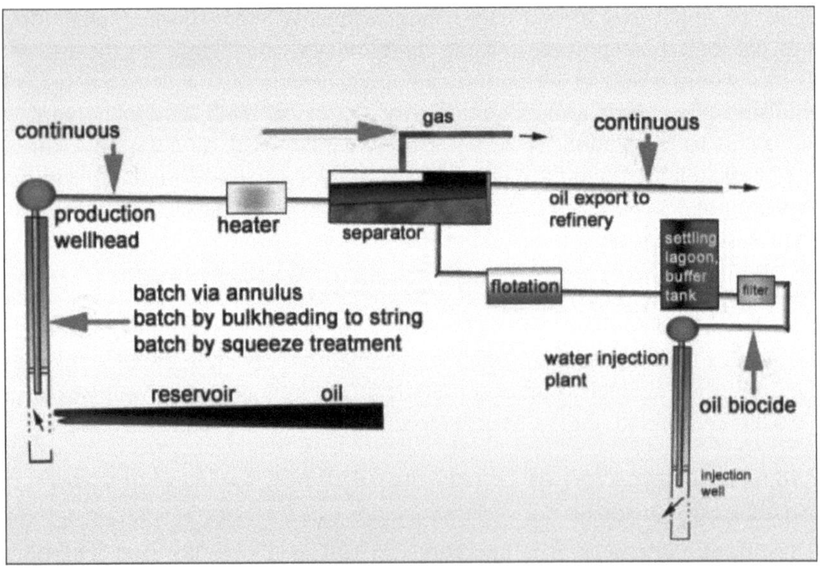

FIGURE 2.49 Typical Corrosion Treatment Areas in a Typical Production System.

a level that is economically justifiable, taking into account plant profitability and planned life, whilst at the same time minimizing environmental and safety risks. Wherever possible, corrosion monitoring should be designed into a plant at an early stage. On-line monitoring should be installed wherever it is known from experience that unacceptable levels of attack are possible and in all critical areas where corrosion can be dangerous or expensive. The position and orientation of the monitoring device are important: for example, if separated water is the most corrosive agent, then bottom-of-line monitoring is needed, but if vapours on the upper surface of a wet gas line are the problem, then top-of-line monitoring is also necessary. In general, it is best to use more than one method of corrosion monitoring as the different forms of corrosion respond differently to each monitoring technique. A number of test methods are available and are described below.

Intrusive Methods – Direct Measure of Corrosion These monitoring devices are inserted into a system and provide direct information on corrosion rate, metal penetration, or fluid corrosivity:

- Coupons (on-line), spool pieces
- Electrical resistance probes (ER)
- Linear polarisation resistance probes (LPR)
- Calipers/intelligent pigs

Intrusive Methods – Indirect Measure of Corrosion These are devices that are inserted into a system and provide information from which a decision can be made about system corrosivity. Examples include:

- Hydrogen probes (on-line)
- Galvanic probes (on-line)
- Potential measurements

Non-Intrusive Methods – Direct Measurement of Corrosion These are methods capable of directly detecting metal penetration, corrosion rate, or fluid corrosivity but without intrusion into the system. Examples include:

- Ultrasonics
- Radiography
- Sidestream monitoring
- Thermography

Non-Intrusive/Indirect Methods These include mainly analytical data from which a decision about system corrosivity can be made. Examples include:

- Acidity
- Oxygen content, iron counts (+Mn)
- Fluid composition analysis

- Microbiological evaluation
- Hydrogen-patch monitors
- Acoustic emission

2.10.7 Microbial Activity and Management

The presence of water in the formation and in produced fluids allows microbial activity to be present. The activity of the sulphate-reducing bacteria in particular generates problems in the oilfield production:

- Sulphate-reducing bacteria in bio-films on metal surfaces results in the production of metal sulphide films and corrosion
- Souring of the produced fluids following contamination of reservoir by injection water

It is extremely difficult to estimate the actual cost of corrosion attributable to the activity of sulphate-reducing bacteria in the offshore oil industry. Figures for the annual costs of overall corrosion are astronomical, and if sulphate-reducing bacteria are involved in only a fraction of this cost then, clearly the activity of these bacteria can have severe economic effects. This section will review the petroleum microbiology and outline the possible strategies to overcome the growth of bacteria. The specific actions applicable to these problems will be detailed in other sections on corrosion control, injection water, and produced water.

2.10.7.1 Sulphate-Reducing Bacteria

Present classification of the sulphate-reducing bacteria constitutes an ever-increasing group of individual organisms differing greatly in both morphology and physiology. Common features are growth only under strictly anaerobic conditions and the ability to activate and reduce sulphate through a variety of pathways to sulphide.

All sulphate-reducing bacteria are strict anaerobes, requiring not only the absence of oxygen for growth, but in addition, a low redox potential of around -100 mV or less. The production of sulphide, however, is by no means a characteristic solely confined to the sulphate-reducing bacteria.

There are a large number of other bacteria that can reduce sulphite, thiosulphate, and elemental sulphur (but not sulphate), via a variety of pathways, to hydrogen-sulphide including *Pseudomonas*, *Xanthamonas*, *Acetobacter*, and *Aeromonas*. These microorganisms may produce slime in industrial waters but are not known to contribute to corrosion through their metabolic processes.

The classification of sulphate bacteria has been difficult because of the use of impure cultures and confusion over the substrate. At the present time, 9 *Genera* including *Desulfovibrio*, *Desulfotomaculum*, and *Desulfomonas* and

25 species have been identified, some of which are spore forming and others that are not.

The different species of sulphate-reducing bacteria are largely unrelated in the conventional taxonomic sense, although they obviously form a broad physiologically and ecologically related group. They are widespread in both aquatic and terrestrial habitats, especially anaerobic mud and sediments of brackish and marine environments. The range of organic carbon sources utilised by these genera include lactate, pyruvate, glycerol, malate, acetate, propionate, butrate, aromatic compounds, and ethanol. Hydrogen can act as an electron donor.

A more recent classification has been proposed that divides the sulphate-reducing bacteria into two groups based on their oxidative and metabolic capabilities.[7] First, those stains and species that carry out a partial oxidation to acetate of a limited range of carbon sources, e.g., lactate; and second, those capable of oxidising a much wider range of carbon sources. This second group can be subdivided into those capable of carrying out partial oxidation of substrates to acetate, and those able to perform complete oxidation to CO_2.

2.10.7.2 Sulphate-Reducing Bacteria Mediated Corrosion

Corrosion due to the activity of sulphate-reducing bacteria is characterised in three ways:

- Metal is pitted rather than evenly corroded
- Iron and steel corrosion products dominated by black iron-sulphides
- Occurs only in anaerobic environments

In some cases, the required oxygen free conditions are produced through the removal of oxygen by overlying aerobic bacteria in the bio-film. Sulphate-reducing bacteria are never found in pure culture in nature, but are always associated with complex microbial communities. Furthermore, sulphate-reducing bacteria mediated corrosion and other non-biological corrosion processes are not mutually exclusive. Thus, the proposed mechanisms of corrosion induced by sulphate-reducing bacteria is based on laboratory studies that are on the whole conducted with pure cultures and so may be far removed from the actual mechanisms operating in nature.

Corrosion mediated solely by sulphate-reducing bacteria occurs around neutral pH in anaerobic environments. Oxygen is not involved and the predominant corrosion product is iron-sulphide. Thus, in order to explain anaerobic corrosion two questions must be answered:

- Why, in the absence of oxygen, is there no hydrogen overpotential?
- What is the role of sulphide in the corrosion process?

For many years the mechanisms[7] developed by von Wolzogen Kuhr and van der Vlugt were used to explain this phenomenon:

Anodic reaction :	$4Fe \rightarrow 4Fe^{2+} + 8e-$
Dissociation of water :	$8H_2O \rightarrow 8H+ + 8OH-$
Cathodic reaction :	$8e- + 8H^+ \rightarrow 8H$
Cathodic depolarisation :	$SO_4{}^{2-} + 8H \rightarrow S^{2-} + 4H_2O$
Corrosion Products :	$Fe^{2+} + S^{2-} \rightarrow FeS$

$$3Fe^{2+} + 6OH^- \rightarrow 3Fe\,(OH)_2$$

$$\text{Overall Reaction}: 4Fe + SO_4 + 4H_2O \rightarrow 3\,Fe(OH)_2 + FeS + 2OH^-$$

The utilisation of cathodic hydrogen by sulphate-reducing bacteria result in cathodic depolarisation, allowing corrosion to proceed.[7] The ability to utilise hydrogen is thought to be due to the possession of the enzyme hydrogenase. Later work, indicated that pure cultures of Desulfovibrio species (hydrogenase positive) can oxidize cathodic hydrogen, due to reduction of redox indicator. Desulfotomaculum orientis (hydrogenase negative), however, was completely inactive.

A range of chemical forms of iron-sulphide can be responsible for a range of predicted corrosion rates. The physical form of the ferrous-sulphide film formed in laboratory experiments was found to be related to the concentration of iron in the bulk phase. In general, low ferrous-ion content leads to the development of a tightly adherent, hard, protective sulphide film, whereas high ferrous-ion content leads to a floccular, corrosive film.

Ferrous-sulphide is cathodic to steel and acts to depolarise the metal surface by the adsorption of atomic hydrogen. Growing *Desulfovibrio desulfuricans* using fumarate as electron acceptor in sulphate-free medium showed the quantitative importance of cathodic depolarisation by ferrous-sulphide. The ability of *Desulfovibrio desulphuricans* to depolarise steel under these conditions was greatly diminished. On addition of chemically prepared ferrous-sulphide it was found that the corrosion rate was proportional to the amount of ferrous-sulphide added. Furthermore, by changing the position of the coupon from vertical to horizontal it was seen that the corrosion rate was also dependent on the degree of contact between the sulphide and the surface; corrosion being greater when the coupon was mounted horizontally. Ferrous-sulphide, however, is not a permanent cathode and thus the maintenance of high corrosion rates is dependent on the removal of hydrogen. This may be achieved by the activity of hydrogenase-positive bacteria.

The idea of cathodic depolarisation by sulphate-reducing bacteria being the cause of bacterial corrosion is, however, still held by some engineers. Nevertheless, it is now widely accepted that the production of sulphide is probably of greater quantitative importance in the overall corrosion process. It has been suggested that the conversion of atomic hydrogen to molecular hydrogen reaction may be poisoned in the presence of sulphide. Thus, there may be a buildup of atomic hydrogen at the metal surface resulting in hydrogen penetration back into the steel. This may result in problems of hydrogen embrittlement and stress-corrosion cracking.

Other experimental data has suggested that cathodic depolarisation by hydrogenase enzymes is an artefact caused by experimental procedures. The cathodic reactant was postulated not to be ferrous-sulphide, but hydrogen-sulphide:

$$H_2S + e^- \rightarrow HS^- + \tfrac{1}{2}H_2$$

The hydrogenase may have a role in driving this reaction to the right by the removal of molecular hydrogen, resulting in the further generation of H_2S. Ferrous-sulphide would also be generated and operate as a cathode.

Alternative mechanisms, which do not involve hydrogenase or sulphide, have also been postulated. High corrosion rates can be observed on metal coupons immersed in spent culture medium from which sulphate-reducing bacteria and sulphide have been removed. This implies that a corrosive metabolite may be involved, the most likely candidate being a volatile phosphorous compound, although this has not been specifically identified.

The presence of elemental sulphur has been reported in several situations where pitting corrosion has been seen. Sulphur is a highly corrosive agent and can be formed by the chemical or biological oxidation of sulphide. It has been reported that only dissolved sulphur is corrosive and, therefore, the proposed mechanism of corrosion is by the production of concentration cells, which is analogous to oxygen concentration cells. Particles of sulphur reacting with water will produce localised areas of low pH that may induce high corrosion rates. The effect of oxidising a sulphide film to produce accelerated corrosion rates can be demonstrated by immersing metal foils in active cultures of sulphate-reducing bacteria. Once a sulphide film has formed, the corrosion rates are found to be relatively low.

These observations agree well with field experience where active populations of sulphate-reducing bacteria have been identified in strictly anaerobic conditions, but with little evidence of bacteria-mediated corrosion. Where oxygen ingress has occurred, however, local concentrations of oxygen have brought about very aggressive conditions resulting in high corrosion rates. It is clear, therefore, that although sulphate-reducing bacteria mediated corrosion is associated with anaerobic conditions, the role of oxygen in the overall corrosion process may be of utmost importance.

2.10.7.3 Anaerobic Corrosion in the Oilfield

In general, the corrosion rates determined in laboratory studies are much lower than those reported in the field. As discussed earlier all the postulated mechanisms of anaerobic corrosion by sulphate-reducing bacteria have been based on the results of pure culture studies. Furthermore, most of these studies have concentrated on growing the bacteria in a planktonic state in the bulk phase. Recent investigations have placed greater emphasis on the role of the bio-film in bacterial corrosion. The bio-films produced in the laboratory over a number of days, however, may not be representative of the natural bio-films that will have developed over several months or even years.

Practical experience has shown that a very diverse population of microorganisms, predominantly bacteria, produces a living film on the surface called a bio-film that rapidly colonizes metal surfaces immersed in marine systems. In the case of metal surfaces exposed to open seawater, macroorganisms, i.e., seaweeds, barnacles, etc., may also contribute to the bio-film. Thus, bio-films may range in thickness from a few microns to metres. Heterogeneity exists both in regards to the organisms present and the physical and chemical nature of the bio-film, i.e., pH, Eh, nutrient levels.

In a system where the bulk phase is aerobic, the development of a bio-film on the metal surface may allow anaerobic environments to be set at the metal surface. These anaerobic environments will be produced in areas where the rate of oxygen consumption by active aerobic organisms is greater than the rate of diffusion of oxygen into the bio-film. Furthermore, the heterogeneity of the bio-film may result in varying degrees of bio-film development over a surface; in some areas, no bio-film may be present, whereas in other areas significant bio-fouling may have taken place. Oxygen concentration cells will, therefore, be set up.

Within the anaerobic environments in the bio-film, sulphate-reducing bacteria can become active with the concomitant production of sulphide. Dependent on the development of the bio-film, sulphide may be produced in patches over the surface or it may form a complete sulphide film. The manner of sulphide production will, therefore, play a major role in any subsequent corrosion.

Local pH concentrations may be set up within the bio-film. The hydrogen ion concentration within the bio-film and at the metal surface may be of great importance in regulating not only bacterial activity in the bio-film but possibly in regulating corrosion rates.

It is evident, therefore, that in order to gain a greater understanding of the mechanisms involved in sulphate-reducing bacteria mediated corrosion, and examination of the natural, undisturbed bio-film must be performed. At present, however, field monitoring is almost solely restricted to the enumeration of bacteria in the bulk phase. It cannot be overstressed that the only method of accurately assessing the potential for bacterially mediated corrosion in a system is by monitoring sessile bacteria in bio-films on the metal surface.

It is clear that no single mechanism can fully explain the corrosion rates seen in the field. In most cases a number of factors will be contributing to the corrosion scene. Probably the most important factor in sulphate-reducing bacteria mediated corrosion, however, is sulphide production within bio-films.

2.10.7.4 Ecology of the Sulphate-Reducing Bacteria

The ecology of bacteria is the science of their interaction with their environment. It encompasses the study of:

- Environmental conditions that permit or favour activity and growth of bacteria
- Effects of activity and growth of bacteria on that environment
- Interactions between bacteria and other organisms, which constitute the ecosystem

As applied to bacteria, ecology and distribution do not necessarily correspond, as, due to their ability to remain dormant, bacteria can often be isolated from habitats, which seem totally unsuited to them and on which they appear to be having no effect. The sulphate-reducing bacteria provide an excellent example of this in that they exhibit almost universal distribution but have limited ecology. Sulphate-reducing bacteria have been isolated from a wide range of different environments including those listed in Table 2.6.

This wide range of environments implies a very great range of adverse conditions within which sulphate-reducing bacteria have adapted to grow.

2.10.7.5 *Environmental Parameters*

Temperature: Sulphate-reducing bacteria that have been isolated can grow at extreme temperatures. Psycrophiles that grow at temperatures as low as minus 5 °C have been isolated from deep-ocean sediments. At the other extreme, thermophiles capable of growth at over 90 °C have been isolated from thermal springs and geysers. In the North Sea, temperatures are fairly constant, changing from around 5-10 °C depending on the season. Within offshore process systems, however, the temperature can vary from that of the seawater around the platform to 25 °C or even higher depending on the system. Water injection systems often include heat exchangers for cooling utility water. This may raise the temperature of the seawater flowing through the system to 20-30 °C. In line with chemical reactions in general, all biological activity increases with temperature.

Pressure: The effect of pressure on bacterial growth may be dependent on the chemical and physical properties of the medium as well as the mechanical forces acting on the bacterial cell. Marine bacteria found in the deep-ocean trenches obviously thrive at pressures greater than 16,000 psi. These

TABLE 2.6 Sulphate Bacteria Environments

Soil	Oil Wells	Fresh Water	Gas Wells
Marine Water	Sulphur deposits	Brackish water	Estuarine mud
Artesian water	Sewage	Hot Springs	Salt Pans
Geothermal Vents	Corroding Iron	Rumen of Sheep	Guts of insects

barophilic bacteria obviously actively resist the effects of high pressures. Pressures over 15,000 psi will denature normal proteins and enzymes. The activity of barophilic or barotolerant strains of sulphate-reducing bacteria in oilfields systems will be discussed in later chapters.

pH: As previously mentioned the pH of the medium is important in bacterial growth. Sulphate-reducing bacteria normally show optimum growth around pH 6.5. Growth of a range of sulphate-reducing bacteria over a range of pH from around 5-9.5 has been commonly observed. The large changes in pH encountered in seawater systems offshore possibly play an important role in overall bacterial activity.

Osmotic Pressure: Sulphate-reducing bacteria have been isolated from a wide range of osmotic conditions, from fresh to highly saline waters. Some *Desulfovibrio spp.* can grow in salt solutions of up to 10% NaCl. It can be seen, therefore, that sulphate-reducing bacteria have adapted to grow over a wide range of temperature, pH, and osmotic pressures. Thus, there will probably be a sulphate-reducing bacteria capable of growth in almost every natural habitat on earth except the most common: an ordinary aerobic environment. Their requirement of low redox potential for growth restricts their activity to reducing environments. Nevertheless, more and more reducing environments are being discovered even in the midst of predominantly aerobic systems, e.g., within bio-films. Their ability to become active and grow in these environments is based on very close nutritional inter-relationships between the sulphate-reducing bacteria and a range of other bacteria in the environment.

2.10.7.6 Sulphate-Reducing Bacteria in Microbial Consortia

As a result of their physiology, sulphate-reducing bacteria are always associated with microbial consortia. Within these consortia they are dependent on other organisms, both for the provision of carbon sources and for suitable environmental conditions for growth, i.e., anaerobiosis. In sulphate-rich environments these consortia normally involve the activities of three broad groups of organisms:

- Aerobic, facultative, and anaerobic heterotrophs
- Hydrogen-producing acetogens
- Sulphate-reducing bacteria

The activity of the heterotrophs results in a number of environmental alterations. First, the activity of aerobic organisms will lower the oxygen concentration eventually leading to the production of anaerobic zones within the consortia. Second, the heterotrophs as a group will initiate the breakdown of polymeric material and the fermentation of hydrolysis products. Under the anaerobic conditions produced, the acetogens utilise the fermentation products, e.g., fatty acids and alcohols, generating acetate and hydrogen that are then available for the sulphate-reducing bacteria. Intermediate fermentation products, predominantly fatty acids, may also be utilised by sulphate-reducing bacteria that act as the terminal oxidisers of organic carbon to CO_2.

In low-sulphate environments such as freshwater lakes, the methanogenic bacteria that produce methane and CO_2 carry out terminal oxidations. Thus, in sulphate-rich systems, sulphide is produced, whilst in low-sulphate systems methane is produced. Bubbles of methane can often be seen rising to the surface in stagnant freshwater ponds due to the activity of methanogenic bacteria in the sediments.

In all offshore systems, problems relating to the activity of sulphate-reducing bacteria are due to the action of microbial consortia of which the sulphate-reducing bacteria are active members. Pure cultures of sulphate-reducing bacteria are never found in nature. Many of the complex cycles that control the balance of nature involve sulphate-reduction processes.

Probably the most common type of consortia where sulphate-reducing bacteria are active in offshore systems is bio-films. The bio-films may be at an interface or on surfaces, e.g., pipewalls, external members, etc. It is these bio-films that are important in the processes of bacteria-mediated corrosion.

2.10.7.7 Control of Sulphate-Reducing Bacteria

For the safe and efficient operation of an oil-producing installation it is necessary to control the growth of the bacteria. There are ranges of chemical types that have shown to be toxic to bacteria but many of these are too toxic and environmentally unacceptable. For oilfield operations the biocide types normally encountered are:

- Chlorine
- Aldehydes
- Quaternary ammonium compounds

Chlorine and chlorine release chemicals include:

- Chlorine gas
- Sodium hypochlorite
- Chloramines
- Chlorinated quanidines
- Chlorinated tripotassium phosphate (iodophors)

Chlorine is believed to penetrate the cell wall where it contacts the enzyme system, and by oxidising -SH groups on the enzyme, it will block the life cycle of the cell. The advantages of this group of chemicals are that they are cheap, effective against a wide spectrum of micro-organisms, and have a short kill time. The disadvantages are that the chemical will oxidise other organic material and additives, therefore, destroying their function and using up the chlorine before it can attack the bacteria. This results in a potential corrosion problem due to the acidic conditions created.

The aldehydes that may be used include formaldehyde and glutaraldehyde. They are believed to operate by penetration of the cell walls and through reaction with free amino groups of the proteins kill the bacteria. They are cheap and are active for a long time but are toxic and the vapour is toxic through inhalation.

Quaternary ammonium compounds (QAC) have the general formula:

$$R_3 - N^+ - CH_3Cl^-$$

The alkyl groups (R-) attached to the nitrogen vary and may not be the same but are typically long chains (C 12-16) that give the molecule surface active properties and are varied to control the solvency of the compound. The positive charge on the nitrogen gives the molecule a charge that attracts it to negatively charged surfaces.

Very little is known about the mode of action of the quaternary ammonium compounds, although the activity may be attributed to the cationic nature of the chemical causing disruption of the cell wall by reaction with the phospholipids of the cells.

The advantages of this group of compounds are that they are relatively low cost and are effective for a long time as they are only depleted by adsorption on surfaces. They need to be carefully chosen to ensure that they are stable and can cause foaming. The application of these compounds will be related to a particular application when SRB control is discussed relative to control of corrosion and preparation of injection water.

Chapter 3

Fundamental Principles of Management of Reservoirs with Sanding Problems

Chapter Outline

3.1 INTRODUCTION

More than 70% of the world's oil and gas reserves are domiciled in clastic unconsolidated formations (sandstone matrices) with a high risk of sand production. The nature of these rocks imposes limitations on the production rate of highly productive fields and makes development of marginal fields uneconomical.

In new deepwater environments and arctic regions reservoirs are:

- Deeper with operations in water depths beyond 3000 m (10000 ft)
- Known to experience high overburden for reservoirs at shallow depths below seabed as obtained in the Gulf of Mexico and Gulf of Guinea
- Further away, requiring possible subsea development, subsea processing, or possible subsea tieback to a central platform process facility
- Complex and characterised by high-pressure and high-temperature (HP-HT) or high-pressure/extremely-low temperature as obtained in arctic, sand, and water production

Developments in Petroleum Science, Vol. 63. http://dx.doi.org/10.1016/B978-0-444-62637-0.00003-8

In such deepwater and arctic environments, multiphase fluid production (gas, oil, and water) is inevitable, and the transport of which is made more complex by the presence of produced sand, hydrates, scale, wax/asphaltene, and of course, inhibitor chemicals.

Sand management is a massive challenge for the petroleum industry as sand production can cause serious problems to well integrity and safety, limit production and affect safety/disposal as well as cause unscheduled shutdowns. Sand production problems are known to increase with the maturity of reservoirs and increase with increases in water breakthrough and production.

The major challenges in the development of the mature, Brownfields and future green fields are in how to:

- Maximise any individual well drilled and field development performance and
- Minimize operational costs and non-productive time and guarantee flow assurance across the entire composite production system, from reservoirs through the wellbore to the topside and flow lines.

Therefore, in developing reservoirs with sanding problems it is important to be proactive in managing the risk of sand production.

Engineers are increasingly being challenged to drill more complex wells and to manage and produce from more complex reservoirs, the majority of which are prone to massive sand production. Therefore, it is important to answer the following key questions as part of any risk assessment and for proactive operational efficiency:

- Is the well going to produce sand or is it producing sand already? This requires knowledge of the reservoir's geomechanical behaviour and the corresponding impact of the reservoir fluids and production.
- When will the sand come in to load the wellbore and what is the corresponding sand production rate? This requires knowledge of sand production prediction methods supported by appropriate geomechanical studies. The fact that a rock fails does not necessary mean it will produce sand. The impact of the drag force imposed by the fluid production is what will eventually lead to sand production. Therefore, a rock needs to fail or be already totally unconsolidated before the production operation leads to sand migration and production.
- Can we prevent or reduce sand production without compromising fluid production and how do we manage a sand producer?
- Why is a production optimisation strategy required to minimise the impact of sand production?
- What is a formation grain size–pore size profile?
- What is the prevailing pore pressure/depletion profile?

- What is the best way to unload the well of debris and sand and what is the multiphase sand transport and cleanup strategy with respect to pipelines and long subsea tiebacks?

The major challenges for any production technology team in the development of sand producing fields especially in deepwater environment are in how to:

- Provide input into the subsea and topside designs in terms of possible quantity and particle size distribution of sand and frequency of sand produced and transported through the wellbore into the subsea and topside facilities
- Optimise the well design to be fit-for-purpose
- Maximize the individual well's production and field development performance
- Effectively manage the sand and multiphase fluid production as well as the facility integrity
- Deploy appropriate a sand management strategy that is fit-for-purpose including a critical evaluation of an appropriate sand control method – surface control vs. subsurface
- Minimise the impact of sand production on the wells and subsea production facilities
- Minimise operational cost and non-productive time and guarantee flow assurance end-to end across the entire composite production system from reservoirs through the wellbore to the topside and flow lines
- Manage the waste disposal including the environmental impact assessment arising thereof

Developing an appropriate solution strategy that is fit-for purpose is extremely important to enable managers of installations and complex fields to improve intervention and production efficiency through continuous process optimisation, reducing non-productive time, and also guaranteeing flow assurance that can reduce lifting costs per barrel. This requires a total integrated sand management strategy (as illustrated in Figure 3.1) that encompasses:

1. Sand Production Rate Prediction, which is the subject of Chapter 4.
2. Sand Monitoring Strategy and dedicated well services to minimise the impact of sand production including deployment of appropriate sand monitoring devices for monitoring surface sand production and efficiency of wellbore cleanup and well test operations. This is covered extensively in Chapter 5.
3. Sand Control Method, whether surface or subsurface, which is the substantive subject of Chapter 6.

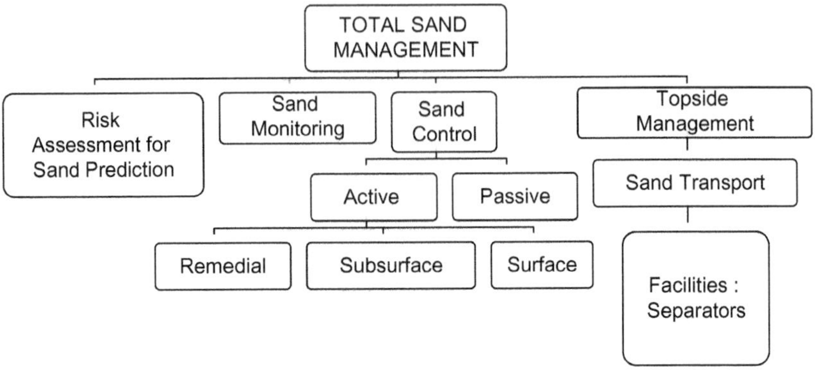

FIGURE 3.1 Integrated Sand Management Nodes.

For wells prone to produce sand, the decision between surface or down-hole sand control critically influences the design, construction, and operation of both wells and facilities. While a risk-averse 'conservative' approach translates into higher initial well cost and lower well potential, a risk-prone 'aggressive' approach could result in high operating cost, significant production deferment, and unsafe operating conditions. The decision between sand control, or not, can swing net present value (NPV) either way by more than 40%.

4. Multiphase sand transport especially in subsea pipelines/tiebacks
5. Topside facility management

3.2 CAUSES OF SAND PRODUCTION

Overall, sand production is a function of rock strength and operational/production conditions. Generally, sand production can be due to the geological condition/structure of the reservoir sand, especially in terms of degree of consolidation, or grain-to-grain cementation of the rock matrix. This is also known as natural causes. Rock failure and corresponding sand production can be due to poor completion and production strategies that encourage sand failure and corresponding sand migration/production.

Thus, the forces that impact sand production are:

- Cementing materials that bond the sand grains together
- Friction between the sand grains coupled with compressive stresses that result in the formation of naturally stable arches
- Cohesion (capillary forces) due to a common fluid phase wetting the sand grains

3.2.1 Natural Causes of Sand Production

For totally unconsolidated reservoir sands there is hardly any or limited weak grain-to-grain cementation. The sand grains are usually held together by osmotic pressure. Such reservoirs are common in the highly permeable Niger Delta Agbada and Akata formations, and the deepwaters of the Gulf of Mexico and Gulf of Guinea. Sand production is usually early, sudden, and can sometimes be unpredictable. Capillary pressure control to minimise impact on osmotic pressure is a way of prolonging onset of production.

3.2.2 Induced Causes of Sand Production

The induced causes of sand production are those due to poor drilling, completion, production, and reservoir management strategies. They include:

a. High production rate beyond the critical limit that induces failure and propagates sand migration
b. Onset of water production that dissolves the grain-grain cementation and breaks down the cohesive osmotic pressure forces
c. Poor drilling, completion, and production strategies including:
- **i.** Landing a horizontal well or perforating deep, close to gas-oil and oil-water contacts
- **ii.** High skin (high mechanical or chemical damage) during drilling, completion, and workover that can result in high drawdown exceeding the compressive strength of the rock matrix
- **iii.** Lack of accurate knowledge of the rock geomechanical and petrophysical properties
- **iv.** Thermal fracture due to water injection cooling, especially in an HP-HT environment
- **v.** Cyclic or shock loading due to repeated start-up and blow-down

3.3 IMPACT OF SAND PRODUCTION

Sand production does have an especially profound impact on the lifting cost of hydrocarbon production. This impact includes:

a. Impact on Well Potential. With the onset of sand production the tendency is to generally choke back with severe loss on deliverability. There could also be severe loss of productivity due to sand deposit in the drains/separators/flowlines. Sudden fill-up of the separator (Figure 3.2) means reduction in available production system capacity that can result in well shut-off and loss of revenue.
b. Integrity Issues
- Downhole completion jewel, well-head systems, and choke rupture due to severe erosion
- Subsea network erosion/corrosion

FIGURE 3.2 Separator Fill-up *(Source: http:/www.intelligent-flow.com[9])*.

c. Process Facility Capacity Reduction due to:
- Sand accumulation in pipework and separator
- Stabilised emulsion in first stage separator, aggravating fines aggregated and settlement in separator base
- Surface equipment damage

d. Reservoir Management:
- Jeopardised produced water re-injection due to severe erosion of injector chokes
- Filtration standard for injection water to avoid particle plugging of reservoir and severe skin effect

3.4 INTEGRATED SAND MANAGEMENT SOLUTIONS

Integrated sand management encompasses all processes, technology, and practices deployed to predict sand production potentials, field processes to prevent formation failure, downhole equipment to prevent formation material from entering the wellbore, 'practices worth emulating' for installing completion to optimise well productivity, monitoring/detection techniques to detect when and where sand is produced, topside equipment for handling produced sand, and workover equipment for performing remedial operations that may be required in the future. Based on the above, an array of issues and solution strategies must be considered if a robust integrated sand management strategy is to be established.

The integrated sand management concept enables the completions engineer to:

- Evaluate the scope for delaying sand control to a later stage of production life (life-cycle economics)
- Determine the need for sand control by evaluating the probability of sand failure
- Evaluate the downhole/sand control requirements and propose an optimum method for controlling sand production

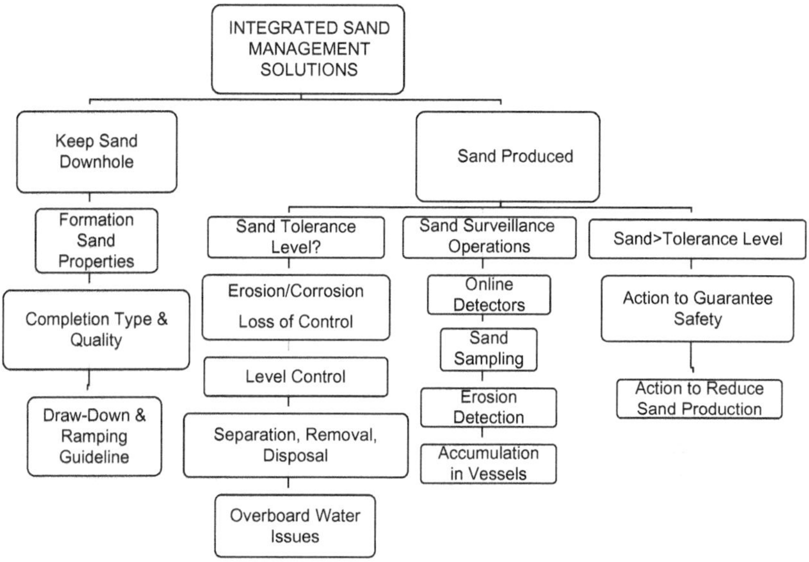

FIGURE 3.3 Integrated Sand Management Solution Options.

- Establish operational limitations (e.g., sand tolerance of topsides, erosional limits, sand detection/monitoring requirements, etc.)

Sand problems require total proactive solutions for good results. The risk of failure can be very high due to the need for proactivity driven by thorough understanding of the critical issues and challenges regardless of whether they concern existing sand producers or new developments. For new developments the bottom line is to try and minimise the risk or avoid sand production, if possible. The main ingredients for integrated sand management are:

- Quality of the well in terms of drilling and completion operations/strategy
- Real-time data analysis
- Evaluation of sanding potential
- Sand grain and produced volume monitoring
- Sand transport/topside management
- Proper sand control strategy
  - Formation grain size profiling
  - Control method
  - Placement strategy
  - Post-performance evaluation

Key questions that drive management strategies and solutions are presented in Table 3.1.

TABLE 3.1 Total Sand Management Questions & Solutions

Questions	Actions & solutions
1. Will well or reservoir produce sand? 2. When, how much, from where?	• Sand Quantification Modelling • Geomechanical Modelling • Extended Well Testing (EWT) • Production Logging Data (PLT Data) • Wireline LOGS Data (GR, FDC, NPH, NMR, etc.)
3. Can the asset team live with the level of sand production? 4. Is the sand being transported? 5. How much erosion can I expect or tolerate 6. What is the regulatory policy about disposal/treatment?	• Transport Modelling • Erosion Modelling • Environmental Impact Assessment (EIA)_ studies • Sand Control Measures • Completion Strategy/Jewels

Consideration of Sand Management Options	• Oriented Perforating
	• Passive Measures
	  • Maximum Sand-free Rate (MSFR)
	  • Water Shut-off Options
	• Bean up or Shut-in
	• Topside Control
	• Downhole Control

Sand Control Option	• Sand Control Completion • Gravelpacks, FracPacks, Screens; etc. • Well Delivery Modelling

For existing sand producers and new developments, the solution must be driven by the proactive need to know:

- Whether the wells or reservoirs will produce sand, when, and the volume of produced sand way ahead of time
- How to maximise production
- How to manage multiphase fluid production
- Choice of downhole sand control vs surface control
- Integrity Issues – Erosion of completion jewels, choke, pipeline facilities
- Topside solids management
- Produced water management (PWRI vs disposal)
- Pipeline solids transport/cleanup
- Flow assurance-emulsion, scale, hydrate prediction and management
- Solids and fluids testing
- What monitoring system to use – acoustic sand detectors (ASD), intrusive sand detectors (ISD), or/and multiphase flow meter (MFM)
- Waste management

Addressing these questions and getting the correct answers requires the following to be fully addressed:

- Detailed Reservoir Characterisation
- Integrated Completion Strategy based on the following key information and strategies:
  - Sand Quantity
  - Sand Control – Subsurface vs Surface
  - Impact on Productivity
  - Sand Transport Efficiency
- Real-Time Data Analysis generated from (LOG, PLT DATA, MDT, WELL TEST (DST, EWT)), etc.
- Knowledge of Hydrocarbon Content obtained from:
  - PVT Data
  - MDT Log
  - Wireline Logs
- Knowledge of Petrophysical Properties obtained from:
  - Logs
  - Core Samples
- Knowledge of Geomechanical Properties based on analysis for:
  - Strength (UCS and TWC data)
  - Maximum Sand Free Rate and Corresponding Drawdown
- Quality of Well and Impact on Well Potential

3.4.1 Need for Sand Management Plan

Executing an appropriate sand management operation must be fit-for-purpose and properly planned. Focus must be on short-term, medium-term, and long-term strategic sand management plans.

3.4.1.1 Short-Term Plan

The short-term plans must address the following:

- Surface sand control through ramp-up or bean down by choke adjustment
- Sand failure diagnosis
- Geomechanical data application to define operational drawdown, depletion, and MSFR
- Establishment and deployment of appropriate sand monitoring systems, either single or hybrid combination of monitoring systems, that include:
  - Acoustic sand monitoring device (ASD)
  - Intrusive sand monitoring device
  - Multiphase flow meter (MFM)
  - Surface or flow line sand sampling
- Address the issues important for minimising the impact of any sand production in the medium term

3.4.1.2 Medium- and Long-Term Plans

In the medium term it will be important to minimise the impact of any sand production by addressing the following:

- Guaranteeing pressure maintenance with a quality-produced water re-injection (PWRI) strategy that addresses:
  - Deoiling and filtration of produced water to remove particulates and to reduce to minimum ppm levels and size so as to guarantee injectivity and minimise formation damage
- Well and topside integrity to minimise the impact of erosion/corrosion on jewels and surface facilities

In the longer term as the well becomes more mature there will be the need to:

- Control well status
- Develop appropriate remedial and mitigation solutions. Some aspects of the remedial options are addressed under in the section 'Sand Control' in Chapter 6.

Chapter 4

Fundamentals of Petrophysics and Geomechanical Aspects of Sand Production Forecast

Chapter Outline

Developments in Petroleum Science, Vol. 63. http://dx.doi.org/10.1016/B978-0-444-62637-0.00004-X

4.1 INTRODUCTION

Sand prediction is a term that refers to all the processes leading to the effective quantification of the propensity of a failed reservoir to produce sand.[10] Any sand prediction estimation starts with an evaluation of the time to rock failure. With this insight, computation of the amount of failed sand is then carried out.

Sand production is one of the major problems facing the oil and gas industry and is common to clastic sedimentary basins throughout the world, affecting thousand of oil and gas fields. Sand production can be sudden and unpredictable. Massive sand production from unconsolidated sandstone can occur as soon as the well is brought on-stream. This may quickly lead to unmanageable problems, in which case sand control methods will be required to continue production. In more consolidated reservoirs, sand production may be restricted to short bursts as the well is beaned up, followed by long periods of relative sand-free production. However, these more consolidated reservoirs are likely to produce sand if a thorough sand management strategy is not established.

Sand management, and the decision whether to exclude or accept sand production, requires an understanding of the mechanisms that cause sanding and the development of a field-validated methodology to predict the critical conditions for sand production. Sand production occurs when the forces on the sand body, induced by fluid flow and corresponding fluid pressure reduction, are greater than the strength of the formation (the loss of mechanical integrity of the formation material). The formation material collapses locally, and the sand fragments are carried into the well bore (separation of solid particles). For most mature reservoirs, water ingress contributes significantly to sand production. Water ingress has destroys the capillary cohesion of the failed sandstone debris and not the intact sandstone. Therefore, in 'friable' sandstone reservoirs, water ingress normally results in transient production of the failed and unremoved debris. In almost all unconsolidated sandstones, water ingress can lead to catastrophic failure.

Sand production is not only a safety hazard for the oil industry but is also a source of revenue loss. Management of sand production requires an accurate knowledge of 'if the reservoir rock will fail', 'when it will fail', and 'how much sand will be produced'. This is routinely referred to as 'sand prediction'. Sand prediction is an element of an integrated sand management strategy that involves the evaluation of risk of rock failure. For an accurate prediction of rock failure and eventual sand production, the strength of the sand at any point is compared with the effective stresses acting over the lifetime of the well (or field). Consequently, rock strengths have to be estimated along with the effective stresses. When the effective stresses exceed the strength, the rock fails and sand production can occur.

The objective of sand production prediction is to couple geomechanical and sanding evaluation models to determine the sanding risk at given flow and drawdown conditions, or the definition of a critical drawdown and/or depletion condition that allows the well to produce sand-free. These models represent key tools in the development of the field and effective sand management and control strategies for the reservoir life cycle.

The key data needed to build and populate the geomechanical model used in sand production evaluations are rock strength, pore pressure, and in situ stress. Rock strength refers to the ability of the rock to withstand the stress environment around the well bore/perforation cavity. The key well bore stress model inputs are pore pressure and the far-field (in situ) reservoir total stresses: vertical stress, maximum horizontal stress, minimum horizontal stress, and their orientation. In porous reservoir rocks it is the effective stresses of the reservoir that are important in terms of sanding evaluations. Effective stress determination requires knowledge of the total stresses, the reservoir pore pressures, and the poro-elastic factor. The original and projected reservoir pressures are required to determine effective stresses on drawdown and depletion, which increase the effective deviatoric (tangential) stresses the rock has to withstand.

Data in the form of drilling/end-of-well reports, completion data, logs, maps, deviation surveys, Drill Stem Test (DST) reports, and core data are used to build and populate the geomechanical model for the interval of interest and to calibrate and test the sand failure evaluation models. The success and accuracy of geomechanical evaluations are highly geared toward the quantity and quality of the data available. The constraints associated with data must be recognized, and it is essential to adopt a pragmatic approach in line with data uncertainties.

The mechanical response of a porous rock to changes in stresses is mainly controlled by its elastic moduli and strengths. These parameters are referred to as the mechanical properties of the rock. Depending on the type of loading (magnitude and duration), the mechanical properties are further divided into static and dynamic properties. Normally, the static mechanical properties of a rock sample are obtained by conducting laboratory measurement on core materials. However, this can be an expensive procedure and it is often the case that core materials are not available. In addition, measurements made on core materials can be unreliable due to core (mechanical) damage effects, namely, the release of in-situ stresses as a result of extracting the cores at the surface. It is therefore more efficient to be able to predict mechanical properties using log data. Since logs are prevalent in the reservoir zone, using log data to predict rock mechanical properties has several other benefits. These benefits, including a continuous presentation of rock mechanical properties versus depth, which is relatively cheaper than coring, are readily available in most old wells. These two techniques are the fundamentals of typical sand prediction models. A pertinent shortcoming with models based on these two techniques is that they are static, i.e., capable of only predicting sanding potential once, more often during drilling. However, it has been established that sand production is a transient phenomenon. This implies that a time function has to be incorporated into the sand prediction model. Moreover, these 'static' models are incapable of quantifying sand production, either at initial conditions or as a function of time. Therefore, it is appropriate to propose a 'dynamic' model that is capable of real-time quantitative sand prediction, from pre-drilling, during production to abandonment of the field. A cradle-to-grave functionality in sand prediction, resulting in a life cycle approach, is proposed and developed in this chapter.

The integrated sand management concept proposes that a sand prediction study be carried out on every well to determine when the formation will fail and what operational procedures can be deployed to delay the onset of sanding. In addition, inaccuracy in measurement of sand produced from reservoirs using available sand monitoring/detection systems has jeopardized the efficiency of existing integrated sand management strategies. This is due to lack of accuracy in existing topside and downhole sand monitoring/detection systems. These systems have introduced errors in terms of volume differences between actual and surface or subsea sand production volume measurements. These differences could be attributed to sand grains that could not be lifted to the surface by the hydrodynamic forces of the fluids.

The first stage of an integrated sand management strategy in the development of any reservoir with potential sanding problems is to determine whether there is any risk of sand production. This is the simplest form of sand prediction. Quite a few approaches to analyse the prediction and control of sand production have been developed. These existing approaches are based on the following:

- Field observations:
  1. **1.** Testing: These are derived from operational events such as drill stem tests (DST) and well tests. The most common technique used is the sequential flow test, in which the well is produced at increasing rates until sand production is observed.
  2. **2.** Measurements: These are actual sand production measurements from process equipment. The most common source of measurement is from the sand detectors/monitors. Also, periodic measurements of produced sand from process separators provide an excellent indication of produced sand volumes.
- Laboratory experiments: These are rock strength measurements carried out on cores. Typical laboratory tests include the thick-walled cylinder, Brazilian tensile, Brinell hardness number, unconfined compressive tests, etc.

- Rules of thumb: These are based on local experience. Many of these rules rely on sonic logs as these give an indication of rock consolidation.
- Theoretical modelling: Early prediction techniques relied on log data to determine the rock strength and, subsequently, the risk of sand production. A number of the techniques used were based on linear/elastic models. In more recent years, these prediction techniques have been made more complex. A few require finite element simulation to function accurately. However, these recent models assume very simplified rock failure criteria and do not consider changes in the failure envelope as the down-hole conditions change.

4.2 INTRODUCTION TO SAND PRODUCTION RATE PREDICTION MECHANISM AND METHODOLOGY

Management of sand production requires an accurate knowledge of 'if the reservoir rock will fail', 'when it will fail', and 'how much sand will be produced'. This is routinely referred to as 'sand prediction'. Sand prediction is an element of an integrated sand management strategy that involves the evaluation of risk of rock failure. For an accurate prediction of rock failure and eventual sand production, the strength of the sand at any point is compared with the effective stresses acting over the lifetime of the well (or field). Consequently, rock strengths have to be estimated along with effective stresses. When the effective stresses exceed the strength, the rock fails and sand production can occur. The total stresses are normally referred to as the vertical or overburden stress, minimum horizontal stress, and maximum horizontal stress (Figure 4.1).

These principal stresses strongly influence rock failure. A commonly used rock failure measurement in laboratory testing of rock strength is the unconfined compressive strength (UCS). With this measurement technique, failure is easily defined. The other is the hollow cylinder or thick wall cylinder tests (TWC).

The magnitudes and orientations of the underground in-situ stresses are of general concern in the oil and gas industry, especially in the realm of rock

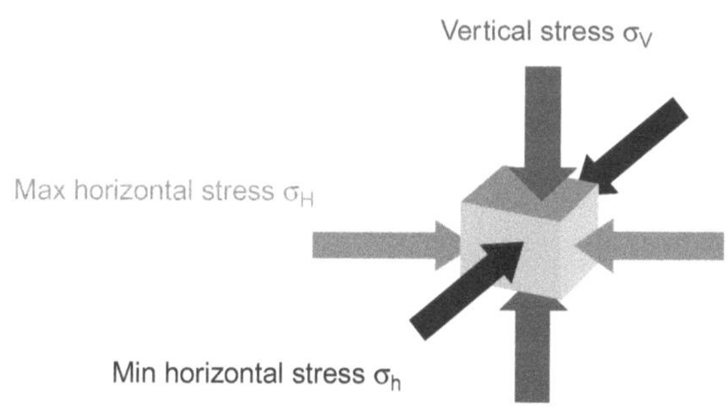

FIGURE 4.1 Rock In-situ Principal Stresses *(Source: Moriwawon[11])*.

failure. Rock failure leads to other unwanted reactions such as sand production and borehole instability.

A known factor with any producing field is the reduction in reservoir pressure. Another important factor is the drawdown generated by the produced fluids. These two factors greatly influence the magnitudes of the principal in-situ stresses. Both depletion and drawdown lead to an increase in the effective in-situ stress. The total overburden stress will remain constant, which means that the effective vertical stress increases. The two horizontal in-situ stresses will also reduce, resulting in an increase in the effective horizontal in-situ stresses. This will reduce the formation strength.

$$\sigma = \sigma_{eff} + \alpha P \tag{4.1}$$

where σ is the total stress, σ_{eff} is the effective stress, P is the pore pressure, and α is the Biot factor.

It is inevitable that the pore pressure will reduce, which will mean the in-situ stress in the rock around the well bore will increase as the rock grains carry the load that was previously supported by pore pressure. The in-situ stress vs. strength relationship is better explained via a typical strength criterion, e.g., the Mohr-Coulomb failure criterion. In the Mohr-Coulomb failure criterion, failure is driven by the differential between $\sigma 1$ and $\sigma 3$ – known as the deviatoric stress. The higher this value, the more likely the rock is to fail. The Mohr-Coulomb failure criterion can be modified to include the UCS (Eq. 4.2). The implication of this is that if the minimum horizontal in-situ stress is known then a limit can be calculated for the maximum shear stress before failure.

$$\sigma_{ucs} = 2C \tan \beta \tag{4.2}$$

where β is the angle between the shear failure plane and the minimum stress direction and C is cohesion.

The pore pressure of a formation may increase, resulting from the generation and/or migration of hydrocarbons. This occurs, for example, in source rocks and overpressured reservoirs. Another process leading to elevated pore pressures is water injection as is the case for the OMEGA fields. An increase of pore pressure has a similar effect as a reduction in vertical effective stress. As a consequence, the horizontal effective stress is reduced as well. As the pressure increases, both the vertical and horizontal effective stresses diminish. The effective stress ratio becomes larger as the pressure increases and may, in reality, be limited by the material yield envelope. However, if the pore pressure is reduced (depleted) in case of hydrocarbon production, the effect is the same as when increasing the vertical effective stress (burial). This leads to an increase in the horizontal effective stresses. Increases in both vertical and horizontal stresses lead to rock instability, which ultimately results in sand production.

Any sand prediction starts with the evaluation of the time to rock failure, and the most important parameter in sanding prediction is porosity. The ultimate aftermath of rock failure is the production of sand. Sand failure does NOT imply sand production. Tensile forces are required to detach the rock grains from the matrix. These tensile forces are commonly exerted by fluid type and fluid flow. Additionally, water influx promotes sand production by reducing the cohesive forces holding the rock grains together. Thus, rock failure MUST occur before sand production.

4.3 CLASSIFICATION OF SAND BY FAILURE TENDENCIES

By analysing formation samples, the engineer can better assess the likelihood of sand production. There are numerous ways to classify sand, with each classification system suited for studying specific characteristics (e.g., depositional environment, chemical composition, particle size, etc.). The well-site geologist traditionally categorized rocks into groups based on visual and manual tests. To achieve a less subjective division for engineering design work on well stimulations, the concept of classifying rocks according to their Brinell hardness number (BHN) was introduced. Hardness tests are used for quick mechanical characterization of (rock) material. The BHN is considered an index property, and a rock classification scheme has been developed that correlates hand inspection with the BHN. The BHN cannot be related to other elastic and plastic parameters in a straightforward manner, but the BHN parameter is useful for characterizing the variations in strength along a core. This technique is quick and cost effective in comparison with other types of testing. Table 4-1 shows the sand classification system.

Table 4.1 provides sample classification guidelines that may also be of use in qualitatively determining sand failure tendency.

Other descriptive classifications are provided such as degree of consolidation, which is used to express increasing difficulty in disaggregration of the sample into individual grains. A rock classification is shown in Table 4.2. Since serious sand production is predominantly associated with the first three consolidation classifications of both Tables 4.2 and 4.3, it is important to discuss them in more detail. When the descriptive terms in the classification tables differ, both terms are listed.

4.3.1 Unconsolidated Sand ($BHN < 2\ kg/mm^2$)

The bulk of sand production problems are found in relatively shallow (<8000 ft), young (Miocene – recent), unconsolidated rocks. Unconsolidated sand can conveniently be subdivided by its degree of natural cohesion as running sand and competent uncemented sand.

Running sand has very small cohesive force or compaction to hold it together. It is difficult to drill through this type of formation without good

TABLE 4.1 Sand Classification System

Descriptive Term	BHN *(kg/mm²)*	Geological Equivalent
Unconsolidated	< 2	No cementing material
Loosely consolidated	2–5	Pieces easily crushed with fingers
Friable	5–10	Pieces crushed when rubbed between fingers
Consolidated	10–30	Pieces can be crushed only with forceps
Hard	> 30	Pieces cannot be broken with forceps
– Low strength	30–50	
– Moderate strength	50–125	
– High strength	> 125	

(Source: Moriwawon[11]).

TABLE 4.2 Whole Core Classification

Rock Type	
Classification by Grain Size	
Unconsolidated	Consolidated
Gravel	Conglomerate
Rubble	Breccia
Sand	Sandstone
Silt	Siltstone
Clay	Shale
Classification by Composition	
Quartz sandstone > 95% Quartz	
Feldspathic sandstone 5–25% Feldspar	
Arkosic sandstone (granite wash) over 25% Feldspar	
Sublithic sandstone 5–25% Rock Fragments	
Lithic sandstone > 25% Rock Fragments	

TABLE 4.3 Sand Consolidation Classification[11]

Descriptive Term	Sample Description
Unconsolidated	Sample disaggregates into individual grains before or after hydrocarbons are removed.
Slightly consolidated	Sample easily disaggregates into individual grains when rubbed between fingers.
Moderately consolidated	Sample disaggregates only after rubbed vigorously between fingers.
Moderately well consolidated	Sample will not disaggregate when rubbed vigorously between fingers. Forceps will disaggregate the sample into individual grains and smaller pieces containing several grains.
Well consolidated	Sample disaggregates with great difficulty into smaller pieces containing several grains using forceps.
Very well consolidated	Sample will not disaggregate with forceps. A hammer disaggregates the sample into small pieces; pieces break across grains.

fluid loss control, as the sand readily collapses into the well bore unless a filter cake is established to provide hydraulic support. Sand production begins immediately with fluid production and sand flows readily with oil, water, or gas. Natural arches tend to collapse immediately as a result of the cohesive forces present. Sand control methods must be used in order to produce these formations.

Wells producing from a running sand type of formation characteristically produce a relatively constant concentration of sand. This can be detected by shakeout tests or by sand monitoring devices. It is often difficult to keep this type of sand out of the well bore long enough to install sand control equipment. Special coring tools must be used to obtain samples of running sand and special gravelpacking procedures must be used to gravelpack the well successfully.

Competent unconsolidated sand results from the increase of in-situ stress with depth that provides the unconsolidated sand with a degree of natural cohesion as a result of internal friction. The well bore stands open through such an interval, even in deviated wells, although the sand is easily washed away during drilling, completion, or production operations. Conventional coring of such sand is difficult. Rubber-sleeve coring is sometimes effective, provided pump rates are controlled. A simple perforated completion can sometimes be used in such a formation, although severe sand production is often experienced if the produced fluid has a high viscosity (e.g., heavy California crudes) or once water breakthrough occurs. Without sand control, some degree of continual sand production can be expected, with sand concentration a function of production rate and drawdown. Competent sand is typical of many Gulf Coast fields.

4.3.2 Loosely Consolidated Sand (BHN 2–5 kg/mm^2) or Slightly Consolidated

Loosely consolidated sand has some cementing agents but only weak, unconfined compressive strengths. A core can sometimes be taken from this type of formation with a conventional core barrel, but it crumbles easily. Rubber-sleeve cores are preferred. An open-hole completion is possible, but the hole tends to collapse and sand itself off when production conditions change. Similarly, perforated completions are initially stable, but as conditions change, cavities, or pockets cave in periodically, especially after sudden rate changes. This causes slugs or clumps of sand to come into the well bore, which may fill the rathole or form bridges in the tubing. Analyses of sand content in produced fluids may show large variations day-to-day as the sand is produced sporadically.

Allowed to continue unchecked, shale and clay beds in the vicinity of the productive sand may also collapse to create a mixture of sandy shale and clays that is difficult to remove from behind a casing. This situation means that any subsequent sand control attempt will have limited production due to the low permeability mixture of clays, shales, and sand around the well bore. Continued sand production from this type of sand formation may lead to collapsed casing after enough sand has been produced to allow the overburden to shift.

4.3.3 Friable Sand (BHN 5–10 kg/mm2) or Moderately Consolidated

The third type of potentially troublesome sand is friable sand, which is well cemented and easily cored. Cores appear strong and competent and do not look as though they would create sand problems. However, under the combination of increasing grain-to-grain stress, erosion, and changes in saturation, the cementation may break down and permit sand production to occur. It is common for this type of formation to produce sand for a few days or weeks after completion and then for sand production to diminish to only a trace or to cease. With significantly reduced pore pressure or water influx into the well, sand production may recur, especially at high production rates where turbulence becomes significant in the near-well-bore area. However, post-failure stabilization effects make it extremely difficult to predict.

4.3.4 Consolidated (BHN 10–30 kg/mm^2) or Moderately Well Consolidated and Harder

This class includes truly well cemented sandstones. Temporary sand production may occur for a very short duration during initial cleanup, choke increases, and following acid stimulation jobs. Otherwise, the chances for intermittent or sustained sand production are very low until much later, after substantial pressure depletion has occurred. The behavioural divisions in the above classifications are somewhat artificial but represent a decreasing probability of sand production.

4.4 GEOMECHANICAL MODEL

The following rock property data are typically required to populate geomechanical models:

- Unconfined compressive strength (UCS, or C_0)
- Thick-wall cylinder strength (TWC)
- Cohesive strength (S_0)
- Friction angle (θ)
- Static Young's modulus (E)
- Static Poisson's ratio (ν)
- Biot poro-elastic factor (α)

Some people believe that direct and definitive data are only available from rock mechanics tests on core. But coring is usually discontinuous, of poor quality and sometimes unobtainable and rock strength data coverage is inherently limited, therefore, the rock strength evaluation for the reservoir is normally based on log indicators calibrated where relevant against viable core data.

4.4.1 In-situ Stresses

Knowledge of undisturbed (initial) principal in-situ stress directions and magnitudes are essential for rock failure evaluations. In fact, it is the starting point of any geomechanical study. Three principal in-situ stresses exist (Figure 4-1), with origins in gravitational loading and tectonics. These are:

- σ_v: vertical stress (also known as overburden stress)
- σ_H: maximum horizontal stress
- σ_h: minimum horizontal stress

The magnitudes and directions of these in-situ stresses depend on the tectonic conditions. These tectonic conditions are directly related to the faulting regimes at play. Based on the Andersonian classification of faulting regimes, three regimes exist, namely:

- Normal faulting.
- Reverse/thrust faulting
- Strike slip/wrench faulting

These principal stresses strongly influence rock failure. A commonly used rock failure measurement in laboratory testing of rock strength is the unconfined compressive strength (UCS). With this measurement technique, failure is easily defined.

The magnitudes and orientations of the underground in-situ stresses are of general concern in the oil and gas industry, especially in the realm of rock failure. Rock failure leads to other unwanted reactions such as sand production and borehole instability.

The orientation of the principal stresses can be determined in several ways. Using four-arm caliper data, or preferably a borehole imaging tool, the stress directions can be traced through consistent orientation of borehole breakouts, and/or drilling induced tensile fractures at the borehole wall. Fractures grow perpendicular to the direction of the minimum horizontal stress. Therefore, the minimum horizontal stress is parallel to the wall of the breakout. These techniques do not permit direct determination of stress magnitudes, but can be used to assess upper and lower limits to stresses. Core-based techniques include anelastic strain recovery (ASR), differential strain analysis (DSA), and acoustic transmission techniques (measuring the stress release induced anisotropy of wave velocities, and the change in anisotropy during reloading). They all require oriented cores, or paleo-magnetic orientation afterward, and do not currently permit reliable estimation of stress magnitudes. In order to provide stress magnitude information, these techniques rely on a good knowledge of the constitutive properties of the rock, and how these are affected by the coring itself (core damage).

The most reliable current method for in-situ stress magnitude determination is to use the average density of the overburden to calculate the vertical stress, and a measure of fracture closure pressure from a fracturing (minifrac or extended leak-off) test in the borehole to estimate the minimum horizontal stress. A standard leak-off test does usually not provide good results, and even if a more elaborate test as suggested above is performed, the reliability depends on the method of measurement (downhole sensors preferable) and interpretation (how to choose closure pressure from the pressure curves). It is, in principle, possible also to estimate the maximum horizontal principal stress from the breakdown pressure, but this is not generally accepted as reliable.

4.4.2 Methods for Minimum Horizontal In-Situ Stress Magnitude Estimation[11]

In some operations, the minimum total horizontal in-situ stress is inferred from extended leak-off data. The horizontal stress magnitudes may vary considerably in different lithologies. Therefore, it is important to make sure that the total stresses determined in shale or mudstones are not assumed to also act in a sandstone or limestone lithology. Some of the techniques used for stress estimation are good in shale, e.g., leak-off tests (LOT – leak-off tests, MLOT – modified leak-off tests and ELOT – extended leak-off tests). However, these tests are not normally performed in reservoir intervals. Conversely, minifrac and microfrac tests are normally performed in reservoir sections and provide good stress data.

Most of the literature on this subject focuses on the passive basin approach to stress estimation and regional correlations to enable a passive basin approach. This is despite the fact that most areas drilled are, in fact, faulted, which is inconsistent with a passive basin approach. These methods predominantly rely on LOT field data, which assumes that the LOT measures the minimum horizontal stress. This is not in fact accurate. The LOT can indicate a pressure anywhere between the minimum horizontal stress and the overburden vertical stress. The lower bound of a large number of LOTs in an area, obtained from vertical wells, does however give an approximation to the minimum horizontal stress in shales. Figure 4.2 illustrates the LOT terminology.

The leak-off pressure (LOP) indicates the point of inflection during pressure build-up and corresponds to fracture initiation. In many cases, the pressure increases beyond LOP due to mud filtrate that plasters and plugs the fracture; this stage reflects stable fracture propagation. The fracture breakdown pressure (FBP) marks the onset of unstable fracture propagation; the pressure reduces accordingly to the fracture propagation pressure (FPP). The minimum in-situ stress is equal to the fracture closure pressure (FCP), inferred from the pressure decline behaviour following shut-in.

A summary of the various methods used in estimating minimum horizontal in-situ stress is given in Table 4.4, and the detailed equations underlying each listed method are given in the following.

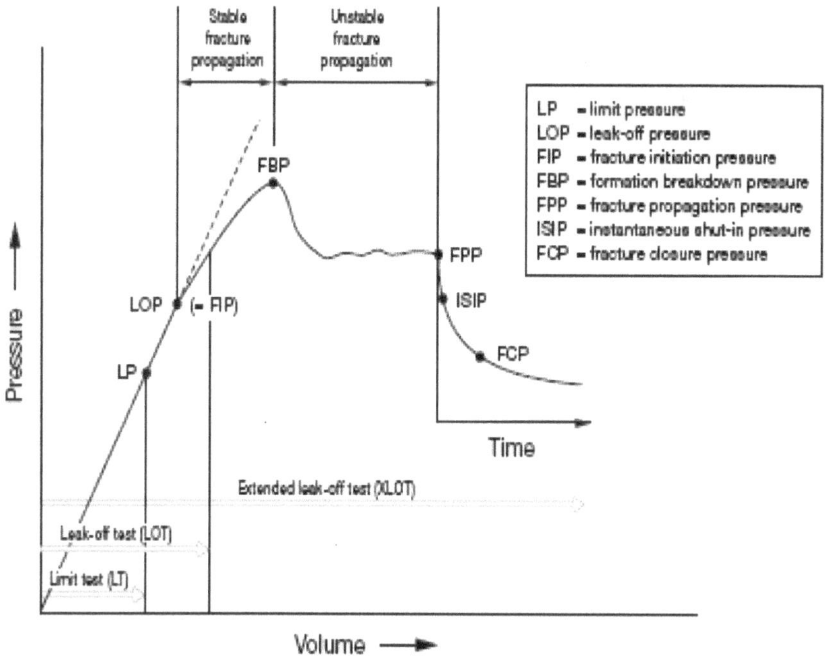

FIGURE 4.2 Leak-off Test Terminology *(Source: Moriwawon[11])*.

TABLE 4.4 Frequently Used Methods for Minimum Horizontal Stress Estimation[11]

Technique	Method	Authors	Applicability
Regional Correlations	LOT & minifrac data	Breckels & van Eekelen.	Shales and mudstones
	Passive basin based approach	Mathews & Kelly, Eaton, Pilkington	GoM
		Eaton & Eaton	Deepwater GoM
Empirical relationships	Holbrook method	Holbrook	GoM Shales only
Theoretical Methods	Passive basin method - elastic	Hubbert and Willis	
		Anderson et al.	
	DCM-based - Passive basin	Heidug et al.	Shales and mudstones
	Daines method	Daines	
Laboratory Core-Based Measurements	Strain relaxation: Anelastic Strain Recovery, Differential Strain Curve Analysis	Ren & Roegiers Voight	Reservoir
Field Measurements	LOT		Shales and mudstones
	Modified LOT		Shales and mudstones
	Extended LOT	Kunze & Steiger	Shales and mudstones
	Minifrac	De Bree and Walters	Reservoir
	Microfrac	Thiercelin & Plumb	Reservoir

Breckels and van Eekelen Model

This is a direct method in the sense that it makes no attempt to predict minimum horizontal in-situ stress through any type of theoretical model. It simply correlates minimum horizontal in-situ stress to some other parameter, like depth. This method attempts to determine the actual minimum stress, not leak-off pressures. Depth is in feet, stress and pressure are in psi, P is the pore pressure, and P_N is normal pressure at that depth. Example equations are:

U.S. Gulf Coast (Also applicable to Gulf of Guinea)

For TVD ≤ 11,500 ft. (4.1)

$$\sigma_h\,(\text{psi}) = 0.197(\text{TVD})^{1.145} + 0.46\,(P - P_N)$$

For TVD > 11,500 ft.

$$\sigma_h(\text{psi}) = 0.197(\text{TVD})1.145 + 0.46(P - P_N) \quad (4.2)$$

With normal pressure gradient = 0.465 psi/ft.

Venezuela

For 5,900 ft. < TVD < 9,200 ft.

$$\sigma_h(\text{psi}) = 0.210\,(\text{TVD})^{1.145} + 0.56\,(P - P_N) \quad (4.3)$$

With normal pressure gradient = 0.433 psi/ft.

Brunei

For TVD < 11.500

$$\sigma_h(\text{psi}) = 0.227(\text{TVD})^{1.145} + 0.49(P - P_N) \quad (4.4)$$

With normal pressure gradient = 0.433 psi/ft.

Mathews and Kelly Model

Matthews and Kelly (1967) assume *K* (what they called the 'matrix stress coefficient') to be a function of the vertical effective stress.

$$\sigma_v = \sigma_v - P_N = 1 \times \text{TVD} - 0.465 \times \text{TVD} = 0.535 \times \text{TVD} \quad (4.5)$$

where σ_v' and σ_v are the effective, and total overburden stresses, respectively, and P_N is normal pore pressure at that depth. The procedure for finding *K* at any depth at any pore pressure is as follows:

- Compute the pore pressure at the depth of interest
- Use a 1 psi/ft overburden stress gradient to find the effective stress σ_v'
- Use Eq. 4.6 to find the depth TVD_N where that effective stress would occur in normal pressure (the equivalent depth):

$$\text{TVD}_N = \frac{\sigma_v'}{0.535} \quad (4.6)$$

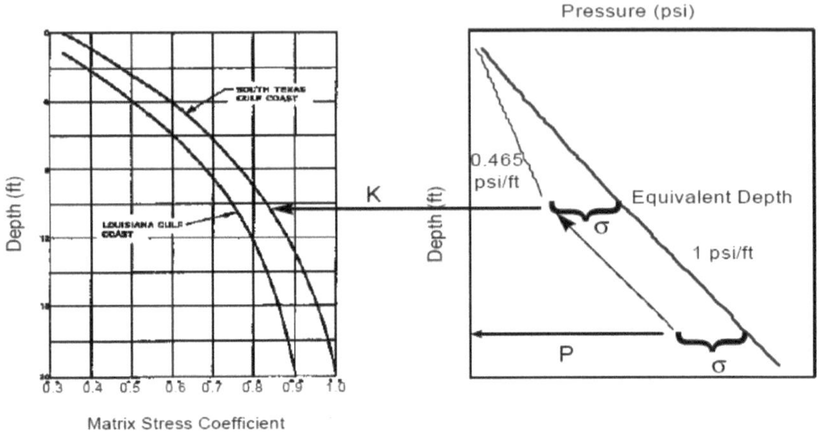

FIGURE 4.3 Mathews and Kelly Method.

Choose the stress ratio at that depth from the appropriate normal trend curve.

Figure 4.3 shows graphically how K is determined.

Eaton Model

To make Eaton's method work, fictitious Poisson's ratios v (Poisson's ratio has no relation to linear poro-elasticity), backed out from leak-off test data must be used to compute K as follows:

$$K = \frac{v}{1 - v} \tag{4.7}$$

where K is assumed (what they called the 'matrix stress coefficient') to be a function of the vertical effective stress. This is done at as many leak-off test points as possible, and then a v vs. depth relation is fit. However, in the absence of leak-off test data, Eaton published two analytical relations for v as a function of depth below mudline (TVD_{BML}):

Gulf Coast

For $0 \leq TVD_{BML} \leq 4{,}999.9$ ft

$$v = -7.5 \times 10^{-9} \times (TVD_{BML})^2 + 8.0214286 \times 10^{-5} \times (TVD_{BML}) + 0.2007142857 \tag{4.8}$$

For 5000 ft $\leq TVD_{BML}$

$$v = -1.7728 \times 10^{-10} \times (TVD_{BML})^2 + 9.4748424 \times 10^{-6} \times (TVD_{BML}) + 0.3724340861 \tag{4.9}$$

Deep Water Gulf of Mexico

For $0 \leq TVD_{BML} \leq 4{,}999.9$ ft

$$\nu = -6.089286 \times 10^{-9} \times (\mathrm{TVD_{BML}})^2 + 5.7875 \times 10^{-5} \times (\mathrm{TVD_{BML}}) + 0.3124642857 \tag{4.10}$$

For 5000 ft $\leq \mathrm{TVD_{BML}}$

$$\nu = -1.882 \times 10^{-10} \mathrm{x}\, (\mathrm{TVD_{BML}})^2 + 7.2947129 \times 10^{-6} \times (\mathrm{TVD_{BML}}) + 0.4260341387 \tag{4.11}$$

Pilkington Model

Pilkington (1978) used stress ratio data obtained from the hybrid models of Matthews and Kelly et al., to come up with an 'average' stress ratio relation for Tertiary basins:

$$\mathrm{K_{Orig}}(1-0.465)^*\mathrm{TVD} = \mathrm{K_{New}}(\mathrm{OBG}-0.465)^*\mathrm{TVD} \tag{4.12}$$

or

$$\mathrm{K_{New}} = \mathrm{K_{Orig}}\frac{0.535}{OBG-0.465} \tag{4.13}$$

where $\mathrm{K_{Orig}}$ is the original stress ratio at that depth, $\mathrm{K_{New}}$ is the new value, and overburden stress gradient (OBG) is the overburden gradient that would be calculated from Eaton's overburden stress relation.

Pilkington further found that his average effective stress ratio distribution could be expressed as the following functions of (OBG) in psi/ft:

For OBG ≤ 0.94

$$\mathrm{K} = 3.9^*\mathrm{OBG} - 2.88 \tag{4.14}$$

For OBG > 0.94

$$\mathrm{K} = 3.2^*\mathrm{OBG} - 2.224 \tag{4.15}$$

Holbrook, Maggiori, and Hensley Model

Holbrook, Maggiori, and Hensley (1995) assume that *K* is related to fractional porosity ϕ by the simple relation:

$$\mathrm{K} = (1-\phi) \tag{4.16}$$

Hubbert and Willis Model

Hubbert and Willis (1957) used the following relation for *K*:

$$\mathrm{K} = \frac{(1-\sin\theta)}{(1+\sin\theta)} \tag{4.17}$$

where θ is the internal friction angle of the rock. In their paper, they assumed $\theta = 30°$, which results in a value of $K=0.33$. Their stress ratio relation represents the theoretical lower bound for K, known in soil mechanics as the coefficient of active stress '*Ka*'. It defines the lowest value the horizontal stress can be without normal faults developing in flat-lying formation. In tectonically relaxed areas, K is typically much larger than Ka.

Anderson, Ingram, and Zanier Model

Anderson, Ingram, and Zanier devised a fracture gradient relation that accounts for lithology changes. As a first step, they replaced the standard Terzaghi effective stress relation:

$$\sigma_{eff} = \sigma - P_o \tag{4.18}$$

With Biot's relation:

$$\sigma_{eff} = \sigma - \alpha P_o \tag{4.19}$$

where σ is the effective stress, σ is the total stress, P_O is the pore pressure, $\alpha = 1 - K/K_G$, K is the bulk modulus of the dry rock, and K_G is the bulk modulus of the rock grains.

Daines Model

Daines (1982) proposed adding a second term to Eaton's effective stress ratio relation:

$$K = \frac{v}{1-v} + \beta \tag{4.20}$$

where β is a lithology-independent parameter that is supposed to account for tectonic effects. β is backed out from leak-off tests using the following relation:

$$\beta = \frac{(FG - PPG)}{(OBG - PPG)} - \frac{v}{1-v} \tag{4.21}$$

where v corresponds to the lithology the LOT was performed in (typically shale), *FG* is the fracture gradient, and *PPG* is the pore pressure gradient.

In reality, Daines 'β' term in most cases is a correction factor that has to be introduced because he used real elastic Poisson's ratios to compute *K*, instead of fictitious ones. However, the bottom line is that he did figure out a way to make Eaton's method work.

$$K_2 - \frac{v_2}{1-v_2} = K_1 - \frac{v_1}{1-v_1} = \beta \tag{4.22}$$

Therefore,

$$K_2 = K_1 + \frac{v_2}{1-v_2} - \frac{v_1}{1-v_1} \tag{4.23}$$

4.4.3 Methods for Maximum Horizontal In-Situ Stress Magnitude Estimation (Table 4.5)

Routinely, the maximum horizontal in-situ stress magnitude is taken either to be equal to the minimum horizontal stress or as the average of the vertical stress and the minimum horizontal stress. Maximum horizontal in-situ stress magnitude evaluation is commonly carried out as follows:

- Minifrac or microfrac in open holes (possible stress tests prior to hydraulic fracturing)
- Determined from induced fractures from borehole images
- Determined from breakout inversion (Qien and Pederson method)

4.5 BASIC THEORY OF ROCK FAILURE

Rock failure along the borehole wall originates from one fundamental cause: removing a cylinder of rock from a continuous rock mass (i.e., drilling a borehole). The original undisturbed rock is now subject to a 3D in-situ stress field caused by the weight of the material above it, and possibly tectonic radial stresses. The total stresses that were previously supported by the rock now have to be re-distributed and borne by the surrounding rock matrix, especially

TABLE 4.5 Frequently used Methods for Maximum Horizontal Stress Estimation

Technique	Method	Authors	Approach
Graphical/Theoretical Methods	Fault bounds	Moos et al.	
	Breakout formation	Zoback et al.	Back calculation
	Stress bounds	Tan et al.	
Theoretical Methods	Induced fracture formation	Numerous	Back calculation
	Breakout inversion	Qien & Pederson	
Field Measurements	Extended LOT		Shales & mudstones
	Minifrac	De Bree and Walters	Reservoir
	Microfrac	Thiercelin & Plumb	Reservoir

in the near-well-bore region. Consequently, the near-well-bore region will show an increased level of stress.

A 'stress' may be represented graphically as a point in stress space whose three axes are those of principal stress. A change in stress may then be represented as a stress path, the locus of the stress point as it moves through stress space. Rock strength may be defined as the point along any stress path at which the specimen ruptures. A strength surface may be constructed to join all stress points at which rupture occurs. In practice, the strength surface is not unique since it may depend on parameters such as stress path, rate of loading, moisture content of the specimens, lithology, and the way in which strength is defined. A strength criterion is an algebraic expression used to describe the locus of a stress point as it travels on the strength surface. Strength criteria are rarely a close fit over the complete strength surface. Usually when one criterion fits well in one failure region, it may be inaccurate when extrapolated into another region.

These forces (per unit area) are:

- Cementing materials that bond the sand grains together
- Friction between the sand grains coupled with compressive stresses that result in the formation of naturally stable arches
- Cohesion (capillary forces) due to a common fluid phase wetting the sand grains

Factors that affect the strength of these forces are:

- Amount, strength, and nature of the sand grain cementation
- Roundness, roughness, sphericity, sorting, and packing of the sand grains, which will determine the intergranular friction
- Degree of compaction, which is generally proportional to the overburden loading (or depth) and inversely proportional to the reservoir pressure
- Composition, density, and viscosity of the fluids surrounding the grains
- Porosity and permeability of the formation

The de-stabilizing actions that can exceed or reduce the bonding forces and lead to possible sand production are:

- Cementing material can be dissolved or eroded by produced fluids, stimulation fluids (e.g., acid treatments), and injectants (e.g., steam).
- Cementing bonds can be broken by changes in the principal stresses caused by initial drilling and cementing operations, compaction due to pressure depletion, excessive pressure drawdown during production, pressure surges due to facility upsets, large choke changes, stimulation procedures, and sudden starts of artificial lift.
- Drag forces from produced or injected fluids can shear intergranular cement bonds and destroy natural arches of unconsolidated sand. This can result from:

1. High fluid viscosities (generally >50 cp), especially in heavy oil wells where viscosities may exceed 1000 cp
2. Turbulence in the pore throats in gas wells and high rate or high Gas-Liquid Ration (GLR) oil wells
3. Formation damage and plugging of perforations and pore throat areas

- Natural arches in unconsolidated sand can also be destroyed by grain slippage resulting from changes in the principal stresses, pressure depletion, excessive pressure drawdowns, or pressure surges.
- Cohesive forces can be eliminated when the wetting phase becomes mobile. For example, in high viscosity California crudes, it is believed that the viscous crude provides formation stability. Sand production begins as the oil decreases in viscosity and becomes more mobile due to steam injection. Similarly, in typical Gulf Coast wells, the connate water provides intergranular cohesion as oil and gas are produced. However, sand production begins at the onset of water production due to the connate water becoming mobile. Also, the injection of miscible fluids (e.g., as in a miscible CO_2 flood) can lower the cohesive forces by affecting the surface tension of wetting fluids.

Conversely, it seems logical that some actions can ultimately lead to an increase in formation stability. For example, compaction may increase stability by increasing cohesive forces as porosity and permeability are reduced. Additionally, the increased intergranular friction can increase the stability of natural arches.

It should be apparent from this discussion that most sand problems occur with young, poorly cemented rocks, especially at shallow depths; in heavy, viscous oil wells; in highly over-pressured and geo-pressured zones as compaction begins to take place; in certain high rate producers; and in reservoirs that are rapidly depleted and in mature reservoirs with water influx. It should also be apparent that sand production will often be rate-sensitive, since this determines the drawdown, drag forces, and in some cases, the water saturation. A maximum sand-free rate can often be established by very slowly increasing the production rate until a trace of sand production is noticed. Immediately reducing the production rate below this point should restore sand-free production. However, this will change with the production conditions and with reservoir depletion.

4.6 ROCK FAILURE MODELS

As previously explained the rock failure model can be divided into four groups; namely:

1. Empirical model behaviour based on field observations: The first attempt to describe the physical behaviour of rock failure is based on field observations. These rely on establishing an empirical correlation between sand production, well data, and field parameters. A study indicated that records of

sand production spanning over a longer period are most valuable for assessing the influence of depletion and water production. Correlations of sand production field data with field and operational parameters increase in accuracy with the number of parameters measured. However, the amount of data required and field specific character increases accordingly.
2. Analytical models: Analytical models are particularly useful for screening study purposes. They can be deployed over a wider range of conditions. They can also offer insight into cause-effect relationships that are often obscured by models of greater mathematical complexity.
3. Numerical models: Numerical models are usually applied to solve specific issues. They offer much more detail about the failure process, which they are also able to describe with greater sophistication than their analytical counterparts. The finite element technique is mostly deployed in this model. The major setback is that a wide range of data input is required to predict accurately the sand production rate. These data are quite difficult to acquire.
4. Probabilistic models: Probabilistic models make use of analytical models and often incorporate numerical models to assess the effect of statistical variation in underlying parameters based on accepted ranges of uncertainty.

4.6.1 Water Influx and Capillary Pressure

Different rock behaviour before and after water breakthrough results from the changes of rock properties, including both deformation properties (e.g., Young's modulus, Poisson's ratio, bulk modulus, etc.) and rock strength properties. While most strength parameters change with water saturation, the frictional angle is altered little or remains unchanged. For different moisture contents the Mohr-Coulomb envelopes are displaced parallel to each other. However, some researchers have found it varies with water saturation (up to 10%) if the rock surface is chemically reacted with water, thus causing a change in the surface smoothness. Much evidence exists to suggest that in addition to reservoir depletion, the other critical parameter leading to sand failure is water breakthrough. However, the action of capillary pressure in providing a degree of inter-granular cohesion is yet another poorly understood facet of the problem at a micro-mechanical scale. The actual sanding potential could be different from model predictions partly because the pressure gradient may be affected by multiphase flow and the capillary pressure may modify the cohesive strength. Additionally, actual sand production may be associated with a critical flow rate that exceeds the resisting force related to the residual cohesion and capillary pressure, rather than the initial rupture of the formation.

4.7 ANALYSIS OF VARIOUS ROCK FAILURE CRITERIA

All failure criteria are based on effective stresses, and quite a range exists, depending on the use of the 1, 2, or 3 principal in-situ stresses. The various rock failure criteria are as follows:

- Mohr-Coulomb failure envelope
- Drucker-Prager
- Weibols and Cook
- Hoek-Brown
- Griffith criterion
- Tresca criterion
- Modified Lade criterion

Figure 4.4 is a graphical representation of all the listed failure criteria in stress space, all of which are extensively discussed below.

4.7.1 Mohr-Coulomb (Rock Strength Evaluation Depends Only on $\sigma_1 = \sigma_v$ and $\sigma_3 = \sigma_h$)

The Mohr-Coulomb criterion states that the maximum shear stress that can be sustained on a plane is the sum of a constant cohesion component and a frictional component proportional to the normal stress acting on that plane.

The Mohr-Coulomb criterion attributes plastic deformation to shear failure caused by an excessive amount of shear stress. This criterion characterizes rock behaviour in terms of a cohesion C and an internal friction angle θ.

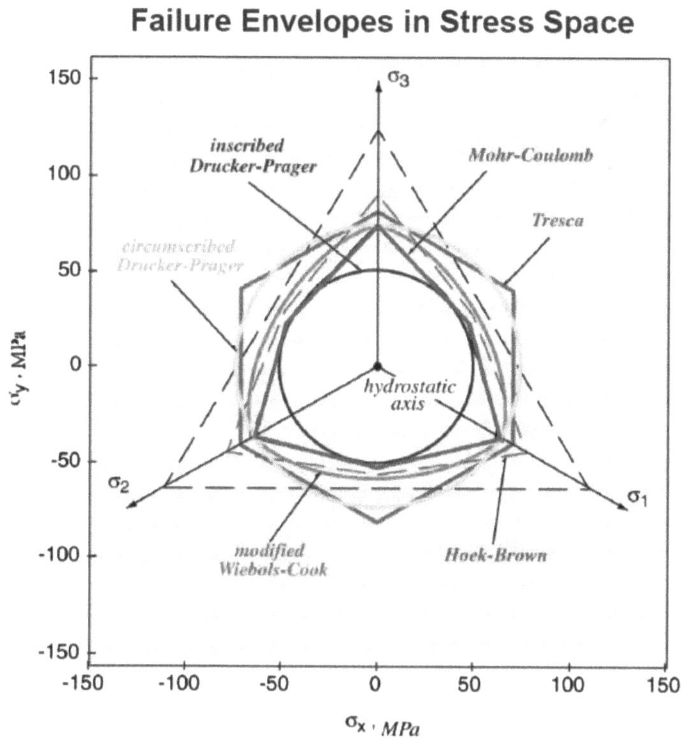

FIGURE 4.4 Failure Envelopes in Stress Space *(Source: Moriwawon[11])*.

It predicts a maximum shear stress increasing with mean normal stress in agreement with routine observation. Mohr-Coulomb provides a reasonable description of rock behaviour in the middle range of compressive stress. The failure envelope is described by:

$$\tau = C + \sigma \tan \theta \tag{4.24}$$

where τ is the shear stress, C is the cohesion, σ is the normal stress, and θ is the angle of internal friction.

The relationship between the angle of internal friction and the fracture plane angle ϕ is given by:

$$\phi = 45^\circ - 0.5\theta \tag{4.25}$$

where θ is the angle between the plane of failure and the direction of the minimum principle stress σ_3.

Mohr-Coulomb is the most commonly used failure criterion because it is the most easily explained in simple physical terms, its relatively few parameters are easily defined from standard laboratory experiments, it is a well-known model within the geo-technical community, and the model is easily implemented in numerical code.

4.7.2 Drucker-Prager (Rock Strength Evaluation Depends on $\sigma_1 = \sigma_V$, $\sigma_2 = \sigma_H$ and $\sigma_3 = \sigma_h$)

A number of theoretical borehole stability models use a different criterion for shear failure called the Drucker-Prager (DP) criterion. The main reason for using stress variants is that the DP criterion is expressed in so-called stress invariants, which make computations easier. The disadvantage of the DP criterion is that the intermediate stress does influence strength significantly, contrary to observed behaviour.

Drucker-Prager is associated with its computer friendliness: one does not need to determine which borehole stress is maximum or minimum. The Drucker-Prager criterion is not appropriate for situations where σ_2 approaches σ_1, such as a vertical borehole.

The shear failure condition according to Drucker-Prager is expressed as follows:

Circumscribed Drucker-Prager Criterion

$$\begin{aligned} & J_2^{1/2} = a + bJ_1 \\ & J_1 = \frac{1}{3}(\sigma_1 + \sigma_2 + \sigma_3);\ a = \frac{\sqrt{3C_o}}{q+2};\ b = \frac{\sqrt{3(q-1)}}{q+2} \\ & J_2^{1/2} \sqrt{\frac{1}{6}\left[(\sigma_1 - \sigma_2)^2 + (\sigma_1 - \sigma_3)^2 (\sigma_2 - \sigma_3)^2\right]} \end{aligned} \tag{4.26}$$

Inscribed Drucker-Prager Criterion

$$J_2^{1/2} = a + bJ_1;\ b = \frac{3\sin\theta}{\sqrt{3\sin^2\theta + 9}}$$
$$a = \frac{3C_o \cos\theta}{2\sqrt{q\sqrt{3}\sin^{2\theta} + 9}};\ \tan\theta = \mu \tag{4.27}$$

where J is the Drucker-Prager yield function, a is a material constant related to the cohesion of the material, b is a material constant related to the internal friction angle of the material, C_o is cohesion, and θ is angle of internal friction.

4.7.3 Weibols and Cook (Rock Strength Evaluation Depends Only on $\sigma_1 = \sigma_v$ and $\sigma_2 = \sigma_H$)

This incorporates the dependence of rock strength on the intermediate principal stress, σ_H, but requires true polyaxial rock strength measurements.

$$J_2^{1/2} = e + fJ_1 + gJ_1^2$$
$$g = \frac{\sqrt{27}}{2C_1 + (q-1)\sigma_3 - C_o}\left[\frac{C_1 + (q-1)\sigma_3 - C_o}{2C_1 + (2q+1)\sigma_3 - C_o} - \frac{q-1}{q+2}\right]$$
$$C_1 = (1 + 0.6\mu)C_o;\ e = \frac{C_o}{\sqrt{3}} - \frac{C_o}{3}f - \frac{C_o^2}{9}g;\ f = \frac{\sqrt{3}(q-1)}{q+2} - \frac{g}{3}[2C_o + (q+2)\sigma_3] \tag{4.28}$$

Where μ is the coefficient of internal friction, e, f, g are material constants, C_o is cohesion, θ is angle of internal friction, and $q = \tan^2\left(\frac{\Pi}{4} + \frac{\theta}{2}\right)$.

4.7.4 Hoek-Brown (Rock Strength Evaluation Depends Only on $\sigma_1 = \sigma_v$ and $\sigma_3 = \sigma_h$)

The Hoek and Brown criterion is like the Mohr-Coulomb criterion in that it is two-dimensional and depends only on knowledge of σ_1 and σ_3. The criterion has been found to work well for most rocks of good to reasonable quality in which the rock mass strength is controlled by tightly interlocking angular rock pieces.

$$\sigma_1 = \sigma_3 + C_o\sqrt{m\frac{\sigma_3}{C_o} + s} \tag{4.29a}$$

where m and s are constants that depend on the properties of the rock and on the extent to which it was broken before being subjected to the failure and C_o is cohesion.

For intact rock, s = 1 and < 1 for previously broken rock. Since sand failure does not necessarily imply sand production, Oluyemi and Oyeneyin's (2010) model for estimating critical drawdown can be used to determine onset of sand production. The model is given as:

$$CDP = \left[\frac{(4A + mC_o) \pm \sqrt{(4A + mC_o)^2 - 16(A^2 - sC_o^2)}}{8}\right] - [n(P_{ri} - P_{rc})] \qquad (4.29b)$$

The equation was coupled to time by taking into consideration reservoir depletion. The dimensionless parameter 'n' represents the ratio of change in critical drawdown with reservoir depletion. When both are equally important, $n = 1$, when depletion is more important, $n > 1$ (2 is usually used), and when depletion has no influence, $n = 0$. This critical drawdown can also be said to be the maximum drawdown for a sand-free production.

Since this study is focused on drilling phase, it is assumed that depletion has no influence at onset of sand production and hence, n will be zero. When depletion starts, Eq. (4.4) can be substituted in the algorithm. The equation for real-time evaluation of critical drawdown at the drilling phase then becomes:

$$CDP = \left[\frac{(4A + mC_o) \pm \sqrt{(4A + mC_o)^2 - 16(A^2 - sC_o^2)}}{8}\right] \qquad (4.29c)$$

where $A = 3\sigma_H - \sigma_h$ and σ_H and σ_h are the maximum and minimum horizontal stresses, respectively. The equation above gives two solutions and the correct one will have to be intuitively determined. The input parameters to the Critical Drawdown Pressure (CDP) equation are thus: maximum and minimum horizontal stresses, Hoek-Brown constants, and unconfined compressive strength (C_o).

4.7.5 Griffith Criterion

Griffith postulates that the failure of brittle materials is due to the growth of pre-existing microcracks. Using this concept, he derived a parabolic constitutive relation between the normal and shear stress that act on the plane parallel to a pre-existing crack and that lead to crack extension i.e., failure. The Griffith criterion is applicable in the tensile to low compressive stress regime. This criterion considers plastic deformation due to tensile crack growth. Failure is expressed in terms of the tensile strength σ_t or the unconfined compressive strength:

$$\tau = 4\sigma^2{}_t + 4\sigma^3 + \sigma_t \qquad (4.30)$$

where σ_t = tensile strength.

4.7.6 Tresca Criterion

The Tresca criterion is a simplified form of the linearized Mohr-Coulomb criterion. It is sometimes called the maximum shear stress criterion. It is commonly used in the description of the strength of metals that have yield strength but do not strengthen with confining pressure.

$$\sigma_1 - \sigma_3 = 2C_o \tag{4.31}$$

where C_o = cohesion.

4.7.7 Modified Lade Criterion (Rock Strength Evaluation Depends Only on $\sigma_1 = \sigma_v$ and $\sigma_2 = \sigma_H$)

This is a three-dimensional strength criterion but requires only two empirical constants, equivalent to C_o and mi, to be determined.

$$\begin{aligned} &\frac{(I_1)^3}{I_3} = 27 + \eta;\ I_1 = (\sigma_1 + S_1) + (\sigma_2 + S_1) + (\sigma_3 + S_1) \\ &I_3 = (\sigma_1 + S_1)(\sigma_2 + S_1)(\sigma_3 + S_1) \\ &S_1 = S_o/\tan\theta;\quad \eta = 4\mu^2 \frac{9\sqrt{\mu^2+1} - 7\mu}{\sqrt{\mu^2+1} - \mu} \end{aligned} \tag{4.32}$$

where l_1 and l_3 are the first and third variants of a stress tensor, S_o is Mohr-Coulomb cohesion, μ is the coefficient of internal friction, and θ is angle of internal friction, $q = \tan^2\left(\frac{\Pi}{4} + \frac{\theta}{2}\right)$.

All the above discussed rock failure criteria are static, only giving an indication of failure at initial conditions. The industry has surpassed such mundane solution to formation failure. The variations of formation strength as a function of production conditions (time) need to be shown, and they should also be able to quantify the expected sand.

4.8 ANALYSIS OF Mohr-COULOMB FAILURE ENVELOPE[12–16]

4.8.1 Mohr Circles

The stress condition at a particular point (at the borehole wall) may for many rocks be graphically displayed using Mohr's stress circle. The Mohr-Coulomb shear failure criterion can also be introduced in this representation.

Consider a cylindrical rock sample inside a triaxial compression chamber. The radial confining pressure σ_3 is applied and the axial load σ_l is increased until failure of the sample occurs. Failure is caused by a critical combination of both shear and normal stresses. The stress σ normal to the plane of failure and the shear stress τ parallel to the plane of failure are given by:

$$\begin{aligned}\sigma &= \frac{1}{2}(\sigma_1+\sigma_3)+\frac{1}{2}(\sigma_1+\sigma_3)\cos 2\theta \\ \tau &= \frac{1}{2}(\sigma_1-\sigma_3)\sin 2\theta\end{aligned} \tag{4.33}$$

where θ is the angle between the plane of failure and the direction of the minimum principle stress σ_3, σ is normal stress, and τ is the shear stress. Figure 4.5 shows a graphical representation of the behaviour of σ and τ as a function of θ, i.e., a circle drawn through σ_3 and σ_1 with its center on the horizontal axis at $(\sigma_1+\sigma_3)/2$, where the radius of the circle is $(\sigma_1-\sigma_3)/2$. This is called the Mohr's circle.

An increase in confining pressure will normally increase the effective stress level of the rock, which is reflected by a shift up to normal effective stress axis. Several triaxial tests at increasing confining pressures will lead to several Mohr's circles. Figure 4-5 shows the result of several triaxial tests at increasing confining pressure. It is seen that when the confining pressure increases, the stress $(\sigma_1+\sigma_3)/2$ (the center of the Mohr circle) as well as the stress difference $(\sigma_1-\sigma_3)/2$ (the radius of the Mohr circle) must be increased to bring the sample to failure.

Stress values below the envelope characterize stable formation and values on or above the envelope indicate formation yielding and failure.

4.8.2 Failure Envelope: Cohesion and Friction Angle

The line drawn tangent to the Mohr circles is known as the Mohr-Coulomb failure envelope (see Figures 4.6 and 4.7). Material points that have stress

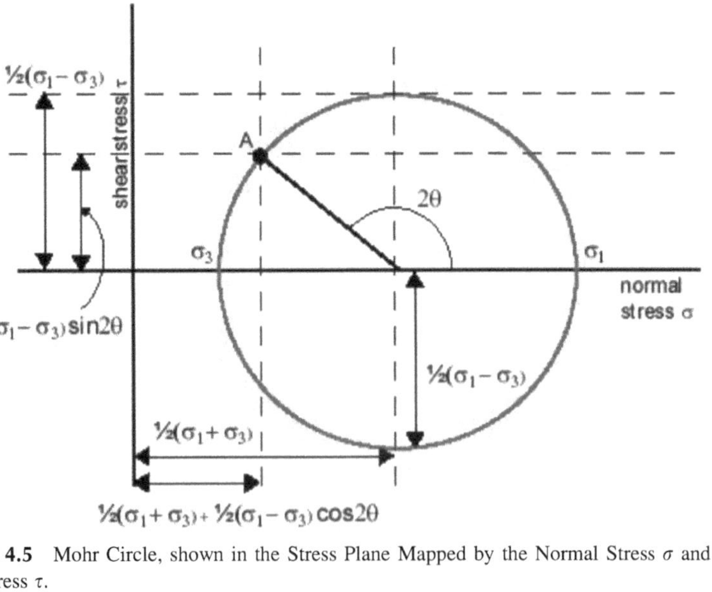

FIGURE 4.5 Mohr Circle, shown in the Stress Plane Mapped by the Normal Stress σ and the Shear Stress τ.

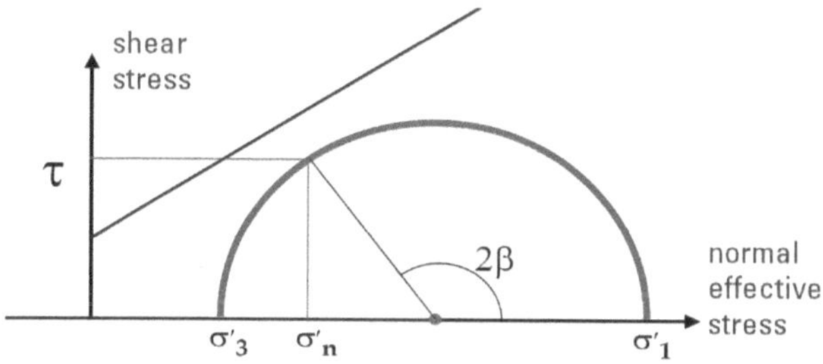

FIGURE 4.6 Graphical Representation of the Mohr-Coulomb Failure Envelope – Signifying Stable Formation.

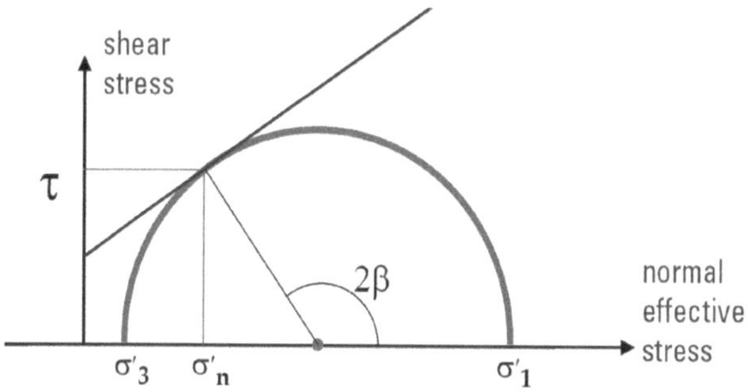

FIGURE 4.7 Graphical Representation of the Mohr-Coulomb Failure Envelope – Signifying Formation Failure.

circles that fall within the envelope are below the point of failure, whereas any combination of stresses outside the envelope will cause failure. The angle that the failure envelope makes with the horizontal normal stress axis is the angle of internal friction θ. The intercept of the envelope on the vertical shear stress axis is the cohesive strength C of the rock.

4.8.3 Determining Formation Shear Strength

There are basically two methods for determining formation strength:

1. **Triaxial testing in the laboratory**. The elastic parameters (Young's modulus and Poisson's ratio) as well as the failure parameters (cohesion and friction angle) can be determined by rock mechanical tests in a triaxial cell. The triaxial test requires rock samples of fair size (typically 1 in. in diameter), which require the availability of suitable core material.
2. **Correlation with well logs and cuttings data**. Suitable core material for rock mechanical testing is rarely available, as coring is expensive and time-consuming. In particular, the sonic log has been found useful to deduce values for Young's modulus, Poisson's ratio, and cohesion. Friction angle has been found to correlate well with the reactive clay content of shale formations, which can be quantified when measuring surface area,

methylene blue test (MBT), or cation exchange capacity (CEC). Refer to the following equations for various relationships between V_c, V_s, E, ν, and CEC.

$$E = \frac{\rho_b V_s^2 (3V_P^2 - 4V_s^2)}{V_P^2 V_s^2}$$
$$\nu = \frac{(V_P^2 - 2V_s^2)}{2(V_P^2 - V_s^2)} \tag{4.34}$$

where E is Young's modulus, ν is Poisson's ratio, V_p is the compressional wave velocity, V_s is the shear wave velocity, and ρ_b is the bulk density.

4.8.4 Modification of the Mohr-Coulomb Failure Envelope

The critical parameters in the Mohr-Coulomb failure envelope are the cohesion (C) and the angle of internal friction θ. Most applications of the Mohr-Coulomb failure envelope assume that cohesion and angle of internal friction are constant for any given sand. These parameters are mostly measured/calculated once, usually at initial conditions. This is grossly misleading. Cohesion and angle of internal friction may vary with respect to the orientation of the bedding or cleavage planes and the minimum principal stress. From previous sections, this work has noted that all the principal in-situ stresses continuously vary with variations in pore pressure. Therefore, if the principal in-situ stresses vary, then the cohesion and angle of internal friction definitely have to vary. However, the expressions for the variance of these parameters are purely empirical in nature and must be determined from laboratory data.

From the preceding discussions, it is clear that sanding is dependent on in-situ stress directions and magnitude, which is dependent on pore pressure. However, the pore pressure is constantly changing as hydrocarbon is being produced. Thus, a realistic sand prediction model should be one that accounts for stress variations as a function of pore pressure depletion (time).

4.8.5 Mohr-Coulomb Failure Envelope Correlations

The Mohr-Coulomb shear failure criterion relates the UCS to the material cohesion (C) and the internal friction angle θ. The material cohesion (C) and the internal friction angle θ are constantly varying with changes in stress. The following correlations detailed attempt to relate the two parameters C and θ with readily available rock/petrophysical parameters. It is stressed that the correlations are intended as a rough guide and should be considered a first approximation. Each correlation presented could be improved upon by more detailed analysis.

Correlation 1: Cohesion vs. Unconfined Compressive Strength and Angle of Internal Friction

Cohesion is the shear resistance at zero normal stress (known in rock mechanics terms as intrinsic rock strength). A direct relationship between the cohesion C (psi) on one hand and the UCS (psi) and the friction angle θ (deg) can be derived if an elastic, ideal-plastic Mohr-Coulomb yield envelope is assumed, namely:

$$C = \frac{\sigma_{UCS}}{2}\left(\frac{1-\sin\theta}{\cos\theta}\right) \tag{4.35}$$

Correlation 2: Angle of Internal Friction vs. Porosity

Angle of internal friction is the angle, θ, between the axis of normal stress and the tangent to the Mohr envelope at a point representing a given failure-stress condition for solid material.

The angle of internal friction can be expressed as a function of porosity (Perkins and Weingarten, 1998[11]):

$$\theta = 58 - 1.35^{*}\phi \tag{4.36}$$

with θ in degrees and ϕ in percentage. The angle of internal friction tends to increase as porosity decreases. Note that the definition of the angle of internal friction depends on the minimum horizontal stress regime.

Correlation 3: Angle of Internal Friction vs. Specific Surface Area

A relationship between the specific surface area S (m^2/g) and the angle of internal friction ϕ (deg) has been developed for cases where this no core material available. The higher the specific surface area, i.e., the higher the amount of smectite present, the lower the shale strength becomes.

The specific surface area can be obtained through a dielectric constant measurement (DCM) as given by the following equation:

$$\log(\theta) = \log(35)\left(1 - \frac{S}{1137}\right) = 1.544 - 1.358 \times 10^{-3} \times S \tag{4.37}$$

Correlation 4: Specific Surface Area vs. Compaction Constant and Porosity

The compaction constant (CC) most effectively controls porosity, which is correlatable to specific surface area (S), which also controls permeability. Therefore, a correlation between CC and SSA will have porosity/permeability embedded in the relation. This is required to enable the user derive S as a function of time as porosity and CC can be easily derived at any time during production. And ultimately, S will enable σ_h to be computed as a function of time.

Compaction Constant Testing (Dataset A)
0.0 – 0.00059
0.0006 – 0.0009
0.0010 – 0.0015

Compaction Constant Calibration (Dataset B)
0.0 – 0.00099
0.0010 – 0.0020

4.9 LOG STRENGTH INDICATORS

There are a number of published and proprietary log-core strength correlations that can be used to develop a continuous rock strength model. There are various examples of such models including the Sarda model[14], Formel[15] model, and Vernik model[16]. These models tend to be constrained by log data availability. Most models involve correlations between unconfined compressive strength (UCS) and logs that are sensitive to rock strength variations – particularly sonic (DTC and/or DTS), density (RHOB), and porosity (PHIT and PHIE). The principal requirement for a log-derived model is that it should provide an accurate description of rock strength in uncored (or unsampled) intervals and wells.

For sand the Sarda model is widely used and considers porosity to exert a basic and implicit control on rock strength – a lower degree of consolidation and cementation (and hence rock strength) is associated with higher porosities. The relationship is described by:

$$\begin{aligned} C_o(MPa) &= 111.5e^{-11.6\phi} \quad (\text{for porosity} > 30\,\%) \\ C_o(MPa) &= 258e^{-9\phi} \quad (\text{for porosity} < 30\,\%) \end{aligned} \tag{4.38}$$

where:

ϕ = porosity (fractional)

The Formel models were developed to describe the main processes occurring in a sedimentary rock during loading. The model was favourably compared with direct rock mechanical tests on cores from over 200 samples in the Norwegian North Sea. The models work best for estimating strengths at low confining stresses (e.g., 0 MPa to 5 MPa).

One model is based on porosity (Model PHI) and the other is based on compressional wave sonic travel time (Model DT):

Model PHI

$$\sigma_{\max} = 43 + 8.0\sigma_c - 0.10\sigma_c^2 - 140\phi + 63\phi^2 - 11\phi\sigma_c$$

Model DT

$$\sigma_{\max} = 140 + 12\sigma_c - 0.10\sigma_c^2 - 2.1\Delta t_c + 0.0083\Delta t_c^2 - 0.063\sigma_c\Delta t_c \tag{4.39}$$

where:

σ_{max} = Maximum failure strength (i.e. $\sigma_{max} = C_0$ when $\sigma_c = 0$), MPa
σ_c = Confining stress, MPa
ϕ = Porosity, fraction
Δt_c = Compressional wave sonic travel time, μsecs/ft

Both models can be used to predict UCS strength if σ_c is set to zero. These models should not be applied outside a porosity range of 20% to 35% and a sonic transit time range of 90 to 140 μsecs/ft.

Vernik et al. cite two models that are referred to as

$$\text{Vernik-1}: UCS = 254(1-2.7\phi)^2 \quad \text{UCS in MPa, porosity } (\phi)\text{, fractional}$$
$$\text{Vernik-2}: UCS = 277\exp(-10\phi) \quad \text{UCS in MPa, porosity } (\phi)\text{, fractional} \tag{4.40}$$

Vernik-1 is applicable for very clean, well-consolidated sandstones with porosities less than 30%, while Vernik-2 is applicable to sandstones with UCS in the range from 300 psi to 50000 psi and porosities from 0.2% to 33%. A modified Vernik model includes a relative shale volume term (Vcl):

$$UCS = (254 - 204Vcl)(1-2.7\phi)^2 \quad \text{UCS in MPa, porosity } (\phi) \text{ fractional} \tag{4.41}$$

A generic (global) relationship can be used to predict TWC from UCS where data are lacking. This is based on large databases of TWC and UCS samples:

$$TWC = 80.884UCS^{0.558} \quad \text{for TWC and UCS in psi} \tag{4.42}$$

In stronger formations (>5000 psi UCS) the UCS/TWC ratio in this database appears constant at around 2, although data acquired in very strong rock shows TWC to UCS ratios approaching 1.5. This is related to the transmissibility of the applied radial stress through the sample, which is much closer to 1 in a hard brittle rock. In weaker formations the ratio increases with reducing strength and can often exceed 6 in intervals with UCS $<\sim$1000 psi. This is due to compaction effects which act to strengthen the sample in TWC tests but which are absent in UCS tests.

Chapter 5

Introduction to Sand and Condition Monitoring Strategies for Asset Integrity

Chapter Outline

5.1 INTRODUCTION

For sand and condition monitoring, there are three main probe systems:

1. *Sand monitoring devices* – There are two versions:
   a. Dedicated erosion probes: These employ intrusive angle-headed probes with a corrosion-resistant alloy-sensing element, typically Inconel 625 (standard).
   b. Non-intrusive monitoring devices
2. *Dedicated corrosion probes*: These employ carbon-steel elements mounted flush with the wall of the pipe/vessel.
3. *Corrosion/erosion probes*: These employ intrusive angle-headed probes with carbon-steel sensing elements. This type of system is very sensitive and can detect and 'fines' in the process.

In this book, we will focus on sand monitoring devices.

Developments in Petroleum Science, Vol. 63. http://dx.doi.org/10.1016/B978-0-444-62637-0.00005-1

5.2 SAND DETECTION/MONITORING

The primary objective of a sand detection/monitoring system is.to provide early warning of sand production from a particular well or transport through the pipework and topside facilities. Early warning implies that:

- the sand detection/monitoring system is more sensitive to sand than the equipment/process it is protecting.
- an alarm is set in a logical location (e.g., control room).
- the device is not sensitive to other factors, and is capable of detecting sand per the design goal (i.e., it can distinguish between 'acceptable' levels of sand production and 'unexpected' levels of sand production).

"From a particular well" implies that the detectors are located upstream of the manifold, or that single wells can be routed for testing. Subsea developments are likely to employ tie-back pipelines (with co-mingled flow) to host topsides, which implies that the sand detector is located subsea at the wellhead or manifold structure.

Sand detection may be achieved in a number of ways including, at the extreme, the unplanned shutdown of a production separator due to high levels of deposited sand (operability problems) to other methods such as the sampling of well-stream fluids. This section focuses on the use of sand detectors (sensors) and monitors that provide early warning of sand production. Sand sampling and measurement of sand accumulation in vessels is also covered.

The three main methods for detecting/monitoring sand are:

1. On-line non-intrusive: ultrasonic, acoustic sensors
2. On-line intrusive: material loss from a probe or element, or possibly the pipe wall
3. Sand sampling from flowlines, pipelines, and process vessels

5.2.1 On-line Non-intrusive Sand Detection/Monitoring

Non-intrusive acoustic detectors operate by detecting noise made by sand grains as they hit the pipe-wall at a bend point in the system. This high-frequency (ultrasonic) pulse is transmitted through the pipe-wall and is detected by an acoustic sensor (transducer) that is mounted on the outside of the pipe wall. The acoustic sensor converts this ultrasonic pulse into an electrical signal (*mV*) that is transmitted, after being amplified/filtered, via a cable to a dedicated computer.

A sensor is installed on a 'bend' in the pipework so the solids in the fluid 'hit' the wall of the pipe causing a vibration. These vibrations, which are of a specific frequency compared to background noise, etc., are filtered and converted to a sand quantity. The power supply for the system can be connected to a total of two sensors to be installed in a safe area, as opposed to sensors that are used in hazardous areas. Cabling between the sensors and the power supply unit (PSU) is supplied, and the PSU is connected to a PC, installed in the control room or safe area lab, with multi-sensor particle-monitoring software installed on it.

Non-intrusive sand detectors require a power supply (to amplify the very small voltages generated by the piezoelectric crystal/transducer) and a data transmission route (to 'handoff' the measured frequency/amplitude information). This is typically achieved by using a cable with two twisted pairs. However, it is understood that this can also be achieved by use of a single twisted cable.

It is recommended that the non-intrusive type of sand detector be installed at a location where the produced sand is expected to impact the pipe-wall. The most common locations are immediately downstream of an elbow or bend in the piping system. At these locations, it is imperative that the detector be located adjacent to the area of impact. In the case of elbows/bends, this will be the outside of the radius.

Non-intrusive sand detectors also detect the noise generated by other noise sources, such as the noise generated by choke valves that sustain a substantial differential pressure (e.g., > 10 barg). Typical guidelines recommend that the acoustic sand detector be located no closer than 4 metres from the choke valve (and not on the first bend downstream of a choke), because the noise generated by the choke can obscure the noise generated by the sand. However, it is difficult to enforce this distance in subsea installations.

If the choke exerts a larger pressure drop such that the flow is choked (ratio of upstream to downstream absolute pressure > 2), then the separation of the sand detector from the choke will need to be extended. If the sand detector is installed upstream of the choke, then noise can also back-propagate through the pipework, and hence, similar separation of the detector from the choke should be maintained (see Figure 5.1). A typical example of an acoustic device footprint is presented in Figure 5.2.

Roxar (http://www.emerson.com)[15] and Clampon (http://www.clampon.com)[16] are the most popular manufacturers of acoustic sand monitoring devices.

Reliable data processing requires the incorporation of sophisticated a Digital Signal Processing (DPS) particle monitor that features complete digitisation and elimination of analogue filters, circuits, and amplifiers. All signal processing is done within the sensor itself so there is no chance for any signal loss or chance of interference from external noise sources. All sensors are alike and interchangeable, which is an advantage if sensors have to be removed/relocated or in case of service.

The DSP's increased processing capacity enables the sensor to combine signals from several frequency ranges when analysing the flow. The sensors are versatile, and the instruments on the market offer two-way communication between sensor and control system. The solution enables future upgrade of the sensor by a simple download of new software. When using digital output from the sensors, the signal from the sensors can be daisy-chained. The sensors

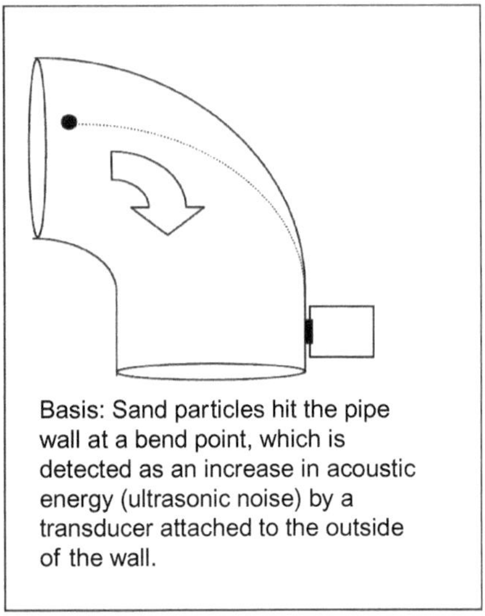

FIGURE 5.1 Techniques of On-line Non-intrusive Sand Detection/Monitoring.

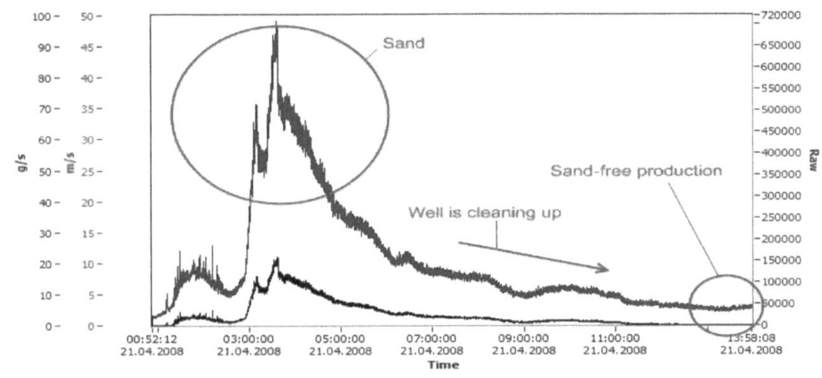

FIGURE 5.2 Example Acoustic Sand Monitor Footprint.

have memory capacity for storing up to 60 days of data. For subsea wells it is essential to locate the sensors close to the subsea well via the use of appropriate subsea funnels (Figure 5.3).

The benefits of acoustic sensors include:

- High sensitivity to sand production
- Instantaneous response to sand production
- Low power consumption
- Ease of installation, no production shutdown, no hot-work permit required
- No temperature dependency effects

FIGURE 5.3 Typical Clampon Subsea Acoustic Sand Monitoring Funnel.[17]

- Reduced risk and cost
- Ideal solution for HP/HT wells
- No pressure-bearing seal failure risk
- Reduced manual handling
- No inherent component/mechanical parts, no fear of wear and tear
- Non-intrusive, acoustic, probes have individual mounting straps and are installed on a bend in the pipework upstream of the choke

The challenges acoustic sensors include:

- Training/experience required for data analysis and interpretation
- System requires elaborate calibration for accurate sand quantification
- Low sand impingement on pipe-wall in low velocity, high viscous flow
- Very sensitive to transient changes in fluid phases

5.2.2 On-line Intrusive Sand Detection/Monitoring

The operating methodology of the intrusive sand detector is akin to that of an electrical resistance (ER) corrosion probe. Both types of detectors make use of elements that are exposed to the flow. However, sand (erosion) probes differ from corrosion probes in that sand probes make use of elements that are made from a material that is resistant to corrosion (whereas corrosion probes use elements that corrode). There are angle and flush probes (Figures 5.4 to 5.6) and the elements are replaceable for sustainability.

Any material loss from the element causes a reduction in the cross-sectional area, which is detected as an increase in the electrical resistance of current flow through that element. In order to compensate for flow-related effects (e.g., temperature) that also influence electrical resistance, an element is also included that is shielded from the erosion damage by placing it on the downstream side of the probe (i.e., it is not eroded by sand). The change in electrical resistance is measured using a Wheatstone/Kelvin bridge so that data is obtained independent of the current used to energise the bridge. Both elements are generally at the same temperature and, hence, any temperature-induced change in electrical resistance can be compensated for by comparing the resistance measured in both elements.

Intrusive sand detectors require a power supply (to energize the bridge) and a data transmission route (to 'hand-off' the measured, temperature-compensated, electrical resistance). This is normally achieved by the use of a cable with two twisted pairs. One pair trickle-charges a battery pack located on the sensing head, while the other pair provides communication for relaying the signals. A battery pack is needed to enable the system to satisfy the requirement of being intrinsically safe because a greater electrical current is required for making the measurement than can be supplied on a continuous basis through the cable. This aspect limits the measurement frequency (to a maximum of approximately one measurement every 20 minutes) to avoid exhausting the battery.

The raw signal for the intrusive system is the probe metal loss. Metal loss from intrusive probes is readily understood, and fulfils the primary requirement of a sand monitoring system. However, to satisfy the secondary objective, the metal loss must be interpreted as a sand quantity. It is recommended that the intrusive type of sand detector be installed at a location

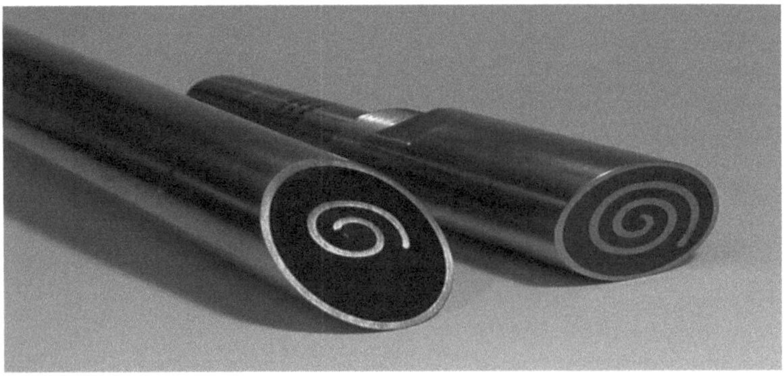

FIGURE 5.4 Angle and Flush Probes.

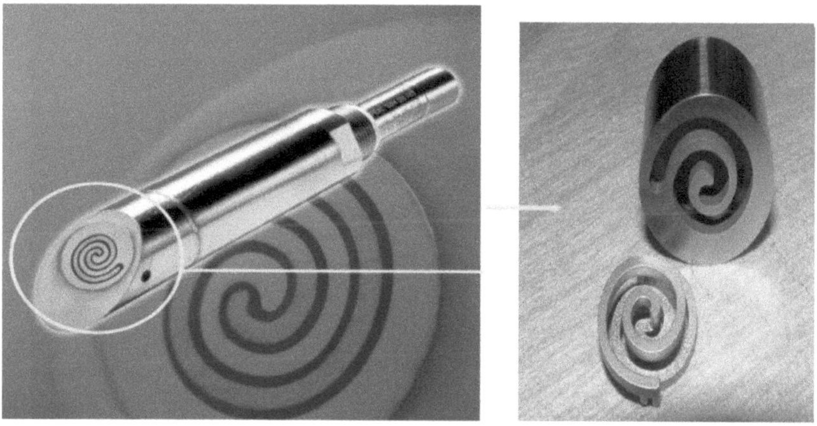

FIGURE 5.5 Angle Probe Showing Replaceable Elements.

FIGURE 5.6 Flush Probe System.

where the produced sand is expected to impact the probe element. Installation downstream of a choke valve is common, but choke valves that become damaged (e.g., by sand) may cause the flow to be directed to one side of the pipe and sand may, therefore, miss the erosion probe. To avoid such scenario, it is common to:

1. Install two probes in a dedicated properly-sized spool piece that can retrofit into the flow line and allow homogenous/heterogeous particle suspension in the flow stream.
2. Probes aligned at right angle to each other to 'capture' particle movement.

5.2.2.1 Commercial Intrusive Sensor for Well Test

Sand monitoring equipment consists of a purpose-built spool piece 1 metre long with mechanical access fittings to install probes. The spool piece should

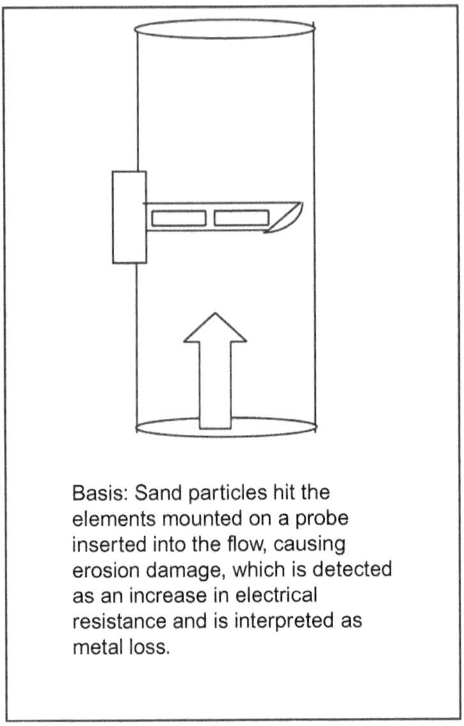

FIGURE 5.7 Techniques of On-line Intrusive Sand Detection/Monitoring.

FIGURE 5.8 On-line Intrusive Detector Shown Mounted on a Horizontal Length of Pipe.

FIGURE 5.9 Example of On-line Intrusive Probe showing the Eroding Elements.[18]

be certified to either 6,000 or 10,000 psi working pressure and have end fittings to mate with existing pipework downstream of the choke manifold. The erosion-monitoring probes should be installed through 2 in. mechanical fittings installed in the spool piece.

Erosion-rate measurements of the sensor are done using high-resolution metal loss technology. Due to the very high-resolution obtained (the resolution is in excess of 100 times that of conventional ER), the results may be used in real time for proactive corrosion/erosion management and system optimisation. Measurements are made and transmitted with temperature data to an acquisition system. The sensor measures the metal loss and temperature of a sample element in the process with a very high level of resolution. This provides the ability to rapidly detect erosion rates and register even the smallest 'sand events' instantaneously.

The intrusive monitoring system has been designed for ease of integration. The output from instruments (metal loss and temperature) is provided in engineering units and tied in directly to the dedicated onboard PC. Basic parameters (erosion rate and temperature) can be converted to erosion rate. Once the data is processed, clear accumulative sand production and sand rates are calculated.

The transmitter consists of an intrinsically safe instrument that is field mounted within 3 m of the intrusive probe. The 24 V DC power is supplied remotely via an isolating barrier and data is transmitted over a separate digital data link to a dedicated PC housed in a 'safe area'. The probe cable is integrated with the instrument and cannot be extended in the field.

A typical spool piece example and data management spread are shown in Figure 5.10.

5.2.2.2 Key Features

- Simple, reliable, and robust
- Instantaneous direct measurement of sand erosion
- Probe can be flush mounted or intrusive into process (Figure 5.11)

FIGURE 5.10 Sand Monitoring System+Spread *(Source: PMAC[19])*.

- Does not require on-site calibration
- Not subject to noise interference
- Sensitive enough to detect fines at low velocity, viscous fluids
- Instantaneous direct measurement of sand erosion
- Probe can be flush mounted or intrusive into process
- Does not require on-site calibration
- Not subject to noise interference
- Sensitive enough to detect fines at low velocity, viscous fluids

Examples of output footprints from a commercial intrusive sand monitoring system and corresponding sand rates are presented in Figures 5.12 and 5.13.

5.2.2.3 Gaps Common with Ultrasonic Sand Detectors/Monitors

Ultrasonic measurement is based on detecting high-frequency sounds from outside a pipe-wall. These sounds are generated either by the flow medium in the pipe or, as in this case, by particles in the medium. The ultrasound is generated by the energy produced when the flow itself or particles impact the inside of the pipe-wall, which thus acts as a membrane that absorbs energy from the particles. A sensor on the outside of the pipe-wall captures this energy in the form of ultrasound. The sensor is tuned or calibrated to pick up sounds within the frequency range produced by whatever we are listening for, i.e., sand or coal particles, liquid, gas, etc. The signals picked up by the sensor are processed electronically by the 'intelligent' part of the sensor before they are sent to the computer, which stores all data and presents them graphically on-screen.

Ultrasonic measurements do not count particles or measure particle size. The measurement data are based on picking up ultrasonic signals from the pipe-wall. It is thus impossible to know, for example, whether a single big

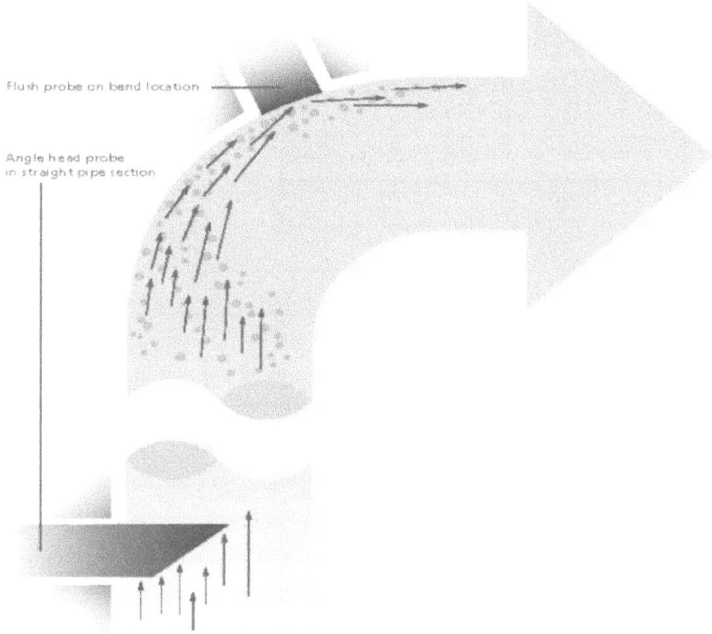

FIGURE 5.11 Installation Options.

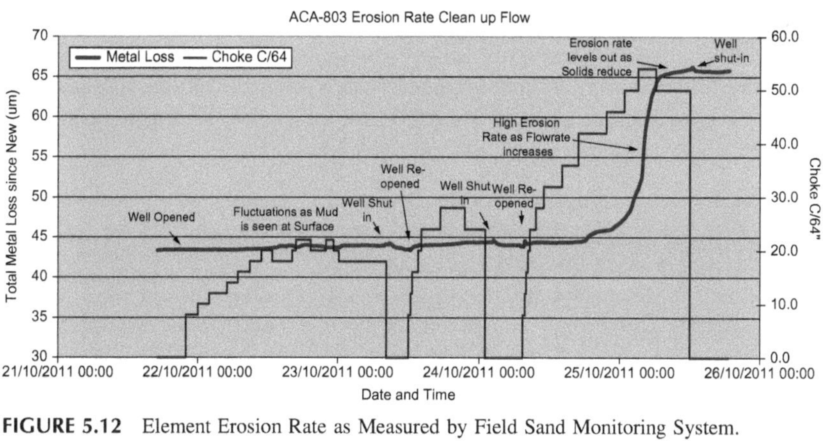

FIGURE 5.12 Element Erosion Rate as Measured by Field Sand Monitoring System.

particle impacts the pipe-wall or the same noise level is due to several small particles. Estimating the size or numbers of particles would require frequencies so low that there would be a danger of picking up other sources of noise. Measurements of this sort are associated with the following sources of uncertainty:

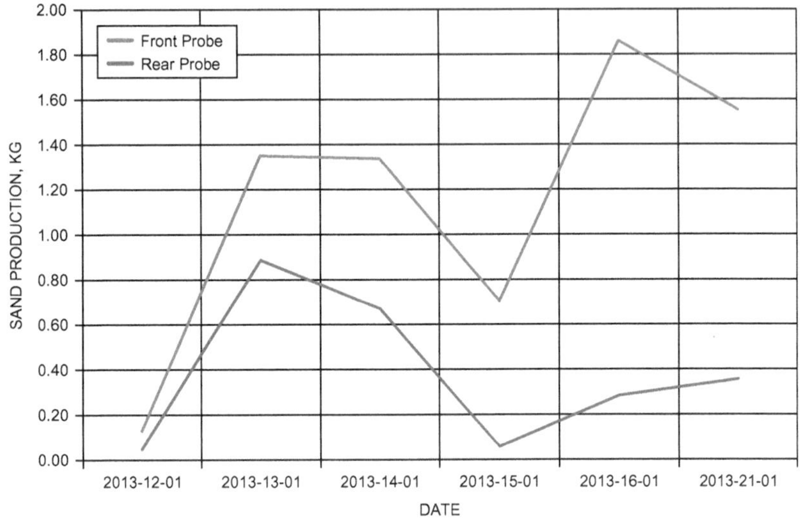

FIGURE 5.13 Typical Sand Rate Output from Monitoring Software.

- Dimensions, density, and surface characteristics of particles
- Impact site of particle relative to sensor location
- Particle velocity and flow velocity
- Distribution of liquid/gas/solids at point of measurement
- Local flow conditions (pipe geometry)
- Other sources of noise

In addition, converting a signal from a pure data value to, for example, grams of particles per second, naturally depends on being able to estimate the velocity of the flow medium, the pipe diameter, etc. Major deviations from correct values may well result in false interpretations of the measurement results.

The gaps highlighted above can be reduced/eliminated by:

- Selecting the appropriate sensor. It must be sensitive to the appropriate frequency range for the particles or energy involved. In this way we can be sure of picking up the correct signals.
- The sensor electronics filters and processes the ultrasonic signals, excluding signals from other sources than particle noise. This minimises the problem of unwanted background noise from other sources.
- Extensive calibration via sand injection tests. Also repeated calibration of measurements with actual field data is highly recommended.

5.2.3 Sand Sampling from Flowlines, Pipelines, and Process Vessels

Sampling is the 'traditional method' of detecting sand and often uses the same equipment and procedure as a sample for water content. Traditionally, a minimal quantity of liquid is extracted from the flowline through a sample point (side-tapping). The associated gas is vented and the liquid is examined for the presence of sand. Since the quantity of sand that can be expected in a sample is very small, and the grain size is often small (50–250 μm), filtration and weighing plus examination under a microscope provides the best analysis route. If the filtration is conducted as the sample is taken, this may restrict the draw rate from the well to a low flow. Filters are also prone to tearing, rendering the sample worthless.

When sand accumulates in vessels, it reduces the volume available for liquids, which reduces the residence time and reduces the efficiency of oil–water separation. Also, the reduced liquid volume renders liquid level control more difficult as the liquid level becomes more sensitive to changes in inflow rate or outflow rate (valve opening). Accumulated sand may also restrict level control instrumentation off-take lines. Finally, sand deposits may lead to under-deposit corrosion of the vessel wall when the fluids are corrosive (water+dissolved CO_2) and the vessel material is not resistant to corrosion (e.g., low-alloy steel).

Historically, visual inspection with the vessel off-line, de-pressurised, and vented was the main method of checking for sand – prior to manual removal of the sand. However, since a number of jet-washing techniques have been devised for removing sand without vessel entry, the requirement to monitor sand accumulation by non-intrusive vessel inspection means has become essential.

Common methods for monitoring sand accumulation in vessels while the vessel is in service are:

- Thermal imaging: Thermal imaging works by viewing the vessel to be cleaned through a camera, which has been designed to pick up heat signatures. The sand and sludge within the vessel will have a difference in temperature to any other substance in the vessel. The camera will pick up this difference in heat, and a profile of the sand and sludge can be built up, because sand accumulations appear as cold regions at the bottom of the vessel. Thermal imaging is not suitable where the vessel is lagged.
- Neutron back-scatter: Deploys neutron back-scatter transmission.
- Gamma transmission: Deploys gamma transmission.
- Ultrasonics: Ultrasonic methods transmit a pulse of acoustic energy through the vessel wall and into the process fluid. The unit then ‘listens’ for a reflected signal. The time taken and nature of that signal indicate the nature of the interface that reflected the sound signal.
- Load cells (weight measurement): Used to monitor the accumulation of sand in the collection vessel beneath de-sanding cyclones.
- Closed Circuit Television (CCTV): Visual inspection for sand in low-pressure water filled vessels can also be accomplished using a camera deployed through an access fitting.

5.3 SAND TRANSPORT

The ideal production condition is to have all sand grains entering the wellbore lifted to the surface. However, various cases exist to reduce the efficient lifting of these sand grains to the surface. These result in problems in the wellbore. These problems are broadly classified as:

- High-velocity problems due to erosion and erosion-enhanced corrosion, particularly in high-velocity, high-turbulence areas in the wellbore and well-control equipment.
- Low-velocity problems due to deposition of sand, e.g., bridging of the tubing, failure of well equipment such as sub-surface safety valves, stuck tools, etc.

In high-rate oil and gas wells, the main problem associated with the presence of sand in the production fluids is erosion. Over-conservative estimates of permissible sand levels in this case will place unnecessary constraints on production (leading, for instance, to proposals that gravel packing should be used to minimize the risk of erosion). More accurate knowledge of the allowable sand level may allow such limitations to be removed without affecting equipment integrity.

For low-rate producers, on the other hand, the main risk is loss of production due to sanding up before remedial action is taken. Accurate knowledge of the combination of sand content and velocity at which continuous sand removal from the well becomes impossible may allow such situations to be foreseen and dealt with before damage is done.

The above discussion further lends credence to the need for sand prediction cum integrated sand management strategy.

Sand transport is an important component enabling advancement from a sand prediction to an integrated sand management capability: It is required to know how much of the failed sand should be transported to the surface and the impact of the produced sand on the topside. Because sand has a greater density than water, oil, and gas, it tends to settle unless it is prevented from doing so by an upward flow. This upward flow can be created by turbulence, such as in horizontal flow, or simply because the flow is upward vertical.

In horizontal flow there are three basic modes of transportation:

- Sand stagnant at the bottom of the pipe (no transport)
- Creeping bed mode, where a stream of sand 'creeps' over the bottom of the pipe
- Suspension, where sand is equally distributed in the flow

For upward vertical flow there are again three modes:

- Sand settles (no transport)
- Sand travels upwards in general but some grains move downward part of the time
- Suspension flow

For downward vertical flow, sand will always travel downward. This can happen in two ways: sand just falls down and suspension flow. When sand settles, it will slowly fill the flowline resulting in an increase in local velocity, until the build-up ceases. When the flowrate increases, the settled sand will enter the stream causing a 'sand storm'. To keep sand in suspension, the flow must be turbulent and the flow velocity must exceed the settling velocity.

In single-phase flow (gas, oil, water), turbulent flow starts to develop when the Reynolds number (velocity/pipe diameter/kinematic viscosity) exceeds 2500 and becomes fully developed above 10000. At a velocity of 1 m/s, gas and water flows are generally turbulent. Oil is more viscous and laminar flow can occur at practical flow velocities. For the sand to stay in suspension in horizontal flow, turbulence must be able to counter the force of gravity on the sand grains. This occurs if the local flow velocities exceed the settling velocity. For this to occur, the average flow velocity must exceed the settling velocity by several times (10 times is a reasonable value). For upward vertical flow, the velocity restrictions are somewhat less severe but turbulence is still required to keep the sand from migrating to the side of pipe where the upward velocities are lowest and falling back. For downward vertical flow, turbulence is again necessary to prevent the sand from migrating to the middle of the pipe where the downward velocity is greatest.

Note that the calculated settling velocity is only an approximation. Sand grains are not perfect spheres and the viscosity is difficult to measure accurately, especially taking size distributions into account. The calculated velocity can easily be off by a factor of 2.

In multiphase flow, the sand becomes usually associated with the liquid phase and follows the liquid flow regime. In horizontal sections, the flow regime has little influence since it is stratified or stratified wavy. In other regimes, sand is normally in suspension. In stratified flow regimes, sand is transported by the liquid only, either in suspension or as a moving bed. In slug and churn flow, sand will be mixed in the liquid but a detection system must cope with alternating exposure to gas and liquid. In mist flow, sand can be suspended in the gas but can also be inside the droplets. In annular flow regimes, sand will travel with the liquid along the pipe-wall. Sand is transported in a different way by each flow regime. It is therefore important to realize that a change in flow regime can have dramatic effects on the signal from a sand detector. For instance, when slug flow occurs, sand can reach a detector mounted in the middle of a pipe but in stratified flow, sand cannot reach this detector. However, even when mounted at the bottom, response may be suspect.

When sand is not transported in suspension mode, we can expect sampling and sand detectors to suffer because the sand that reaches the detector (if it does so at all) will not be representative of what is in the flow. The detector response will become very dependent on flow velocity because a little more velocity can bring more sand into the flow. Also, the heaviest grains will settle on the bottom and only the smaller ones will reach the detector or the sampling probe.

If the sampling velocity is too large, additional small grains are sucked in. A sand detector that is based on grains hitting an object (both acoustic and erosion probes) is an extreme example of a sampling probe with a sampling velocity of 0 m/s. These probes favour the larger grains and more so if the velocity is lower. This problem also becomes more pronounced at higher viscosities.

5.4 TOPSIDE MANAGEMENT/HANDLING (FIGURE 5.14)

Sand and other solids entrained in reservoir fluids can cause varying problems at the topside. For example, sand deposition in vessels leads to loss of operating volume and can also cause under-deposit corrosion. At higher velocities, usually associated with fluids with a higher gas–liquid ratio (GLR), erosion caused by solid particles impinging on pipes, fittings, and valves can also lead to integrity implications. Therefore, it is important to be able to firstly quantify the effect of sand on topsides wherever possible, which in turn will allow an integrated sand management plan to be put in place to mitigate these effects, as appropriate.

Although the impact of sand production on topsides, whether it is choke erosion or deposition in the production separators, can be assessed in isolation, it is essential to adopt an integrated approach to sand management. This is because it is possible to mitigate sand production, or its effects, either

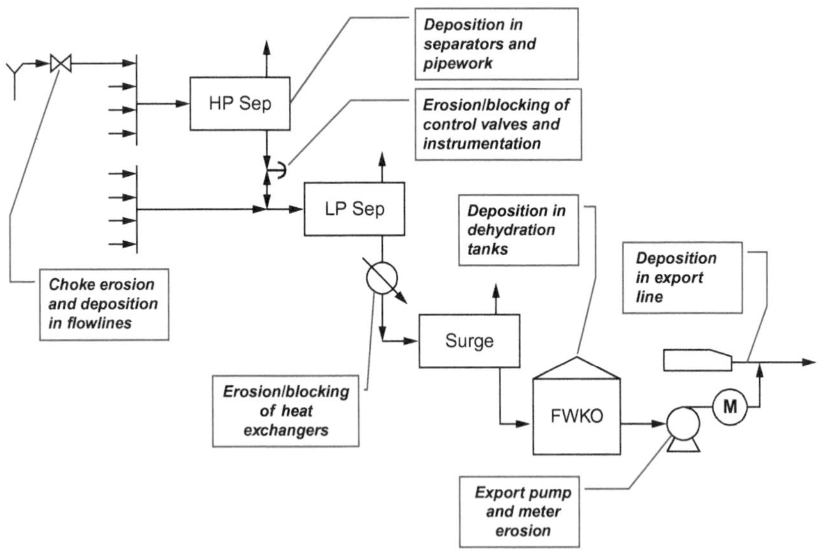

FIGURE 5.14 Example of Typical Sand Production Scenarios. HP = High Pressure; LP = Low Pressure; FWKO = Free Water Knock Off *(Source: Moriwawon[11])*.

down-hole or topside. Both options have associated advantages and disadvantages that must be considered in the evaluation stage, and neither provides a total solution. For example, sand screens do not prevent all sand production, and over time, performance can deteriorate. Primarily, they impair inflow into the well bore, which defers production. Likewise, desander cyclones incur a pressure drop (deferment) and also do not achieve a perfect separation. Because the specific items of equipment do not guarantee a single-point solution, it is important that the completions engineer is aware of the limitations, risks, and tradeoffs associated with using each technology.

Behaviour of sand in production equipment, e.g., its settling and erosion rate, depends on characteristics such as particle size distribution, density, shape, and concentration. To design sand-tolerant topsides, it is important to be able to predict where the sand will deposit. If sand deposits where it is undesirable, e.g., in the flowlines, the pressure drop along the line can increase and under-deposit corrosion can occur. There is also the risk of stuck pigs during cleaning operations. Within separators, sand deposits can reduce effective residence times, resulting in poor separation performance. This can be a major source of deferment.

For most topsides, sand is most likely to deposit in the primary gas–liquid separators due to the relatively low velocities there. Some smaller particles typically carry over with either the hydrocarbon or the water phase, and can settle out in surge vessels or free-water knock-out vessels (FWKO) where the residence time is very high.

Unwanted sand deposits have to be first detected and then removed. Removal can be a difficult process, involving lengthy field shutdowns with associated flushing and cleaning operations. If potential sand deposits can be identified and accounted for in the design phase, automated flushing systems can be installed up-front thereby preventing possible build-up of sand deposits.

Sand deposition in flowlines is most likely in mature assets due to failed or sub-optimal performance of sub-surface completions, low production rates (low velocities), and higher water cut (lower viscosity of fluid). Potential harmful effects of sand deposition in flowlines and pipework include:

- Reduction in pipework and flowline capacity, leading to increased pressure drop and under-deposit corrosion (as the sand allows formation of a corrosive cell)
- Source of erosive material if operating conditions change suddenly, creating a local 'sand storm'
- Potential to cause stuck pigs during flowline/pipeline cleaning operations
- Slugs that have a tendency to choke equipment, resulting in very high erosion of internals in a short time
- Blockage of instrumentation tapings, leading to incorrect process data
- Accumulation of associated trace radioactive scale

Chapter 6

Sand Control Completion Strategy

Chapter Outline

Developments in Petroleum Science, Vol. 63. http://dx.doi.org/10.1016/B978-0-444-62637-0.00006-3

6.1 INTRODUCTION

Sanding in well/reservoir is considered a life-cycle problem that changes over time due to constraints imposed by operational factors of exploration, drilling, production, and stimulation. Predictions made prior to drilling a well cannot therefore be said to be valid during or after drilling and during production operations.

Establishment of a sand management strategy is a logical follow-up to sand production rate prediction. Real-time prediction of sanding potential holds a lot of promise for the industry, especially in relation to the economically wise decision regarding sand control and type of sand control to be adopted, and will become easy for any reservoir management teams. Decisions regarding completion type and design, screen deployment, gravelpack design, and topside management can also be made much easier by developing accurate knowledge of sand production rate.

Real-time sand prediction is also bound to offer a great potential for optimized drilling and production, providing safety and efficiency. The overall impact of this is reduced life-cycle costs and improved field economics.

With the sanding prediction in place the next step is to establish appropriate sand management philosophy that limits or excludes sand production. This falls into two major categories:

1. Passive sand control involving production rate limitation below the critical sand production rate, which implies operating rate limit wells.
2. Active sand control covering three methods, namely:
   a. Sandface downhole control
   b. Remedial sand control
   c. Surface sand control with solids control facilities

6.2 DEFINITION/OBJECTIVES OF SAND CONTROL

Sand control can be said to represent all the procedures/practices, technology, and installations by which sand coming through or with the potential to come through with the produced fluid from a failed clastic reservoir are either contained or stabilized through consolidation within the reservoir.

The specific objectives and drivers for any sand control completion are to:

- Maximise productivity
- Minimise impairment of the reservoir
- Minimise load-bearing sand and fines migration/massive sand intrusion

These are key requirements for any successful sand control technique.

6.3 SAND CONTROL METHODS

There are two main sand control categories: Passive and active sand control methods.

1 Passive Sand Control

Passive sand control methods are those methods that mitigate against sand migration and production. A typical example besides the adoption of best practices is the restriction of production on the basis of the knowledge of anticipated failure time or maximum drawdown that can cause failed sand fluidisation. The overall goal of sand control is exclusion of failed sand likely to be produced with the reservoir fluids.

Passive sand control methods are somewhat simple but can be difficult to manage. Restriction of production to mitigate sand production in particular requires thorough knowledge of the geomechanical properties of the rock and the corresponding interaction with the well and its operating window. The sand control problem lies more with the active sand control methods.

2 Active Sand Control

Active sand control methods can be broadly classified as subsurface/sandface, surface, and remedial sand control methods. Sandface sand control methods generally rely on a downhole filter medium that filters the formation sand grains while allowing fluids to flow from or into the formation depending on if the well is a producer or an injector. There are a significant number of them on the market. Active sand control methods are classified as 'downhole screens', 'gravelpack', and the screenless completions ('frac and pack', ResSHUNT, ResQ, and 'chemical consolidation').

Surface sand control methods rely on the use of surface filters installed at the wellhead or as part of the topside central process facilities to remove produced solids. These include strainer units, desanders, and membranes systems. Remedial sand control represents all methods/techniques required to repair/workover failed downhole sandface sand control completions.

Selection of a fit-for-purpose control method depends on the gathering and analysis of key reservoir and well data as well as rigorous laboratory analysis that do add to well costs. The process is elaborate and the success of it depends on reservoir quality index and operating conditions. This is why different companies have adopted different strategies with mixed results. The key decision factors are:

- Economics
- Rate/recovery/facilities
- Completion
- Subsea vs. platform
- HSE and regulatory restrictions
- Results of risk assessment

The risk assessment criteria include:

- Well-operating envelopes
- Completion longevity/reliability

- Application risks
- Operational risks
- Intervention philosophy
- Reservoir management

6.4 DOWNHOLE SAND CONTROL

Active downhole sand control methods include:

- Liner Systems
  - Slotted liners
  - Predrilled liners
  - Expandable slotted/predrilled liners
- Screen systems including expandables: Special application in extended reach (ER) and horizontal wells
  - Prone to mechanical damage
  - Not recommended for running barefoot in gas wells due to high erosion
- Gravelpacking: Generally regarded as the cheapest and most reliable method
- Frac packing
- Screenless completions. These include:
  - Chemical consolidation: Limitations on completion interval
  - Combination method: Involves the use of resin-coated sand
  - Frac and Pack

  Frac and packs are internal gravelpacks where fractures in the formation have been created under controlled conditions and packed with gravelpack sand (or proppant) for sand control. In this method of sand control, an extra wide, proppant-filled hydraulic fracture is created. The extra width is generated in an additional stage subsequent to the conventional hydraulic fracturing treatment. After developing the required fracture length (frac), the fracture screens out at the tip (tip-screen-out, or TSO) and is further inflated and filled with a proppant (pack). The frac and pack treatment is used for bypassing near-well-bore damage and to increase communication with laminated and medium-to-high permeability reservoirs. One of the promising applications is in combination with gravelpack completions, which typically show substantial near-well-bore damage. The optimum stimulation technique for a specific application should be selected considering all aspects involved such as reservoir properties, type of formation, type of damage, operational cost, and expected benefits.
  - Oriented Perforating
- In weakly consolidated sand and formations with large stress contrasts created by tectonic environments, oriented perforations are those that are utilised to target minimum stress directions. They generally reduce the formation failure envelope. This chapter focuses specifically on screen systems and gravelpacking sand control completions. Proper selection and

design of these sand control methods depend on accurate knowledge of the formation sand and well bore solids textural properties.

6.5 FORMATION SAMPLING AND TEXTURAL PROPERTIES ANALYSIS

This Chapter focuses on the process of textural properties analysis includes:

1. Formation sampling to obtain representative samples of the formation sand
2. Detailed sieve analysis
3. Textural characterisation for grain size/size distribution and shape/angularity

6.5.1 Formation Sampling

The best ways to obtain representative samples are via:

1. Real-time grain-size profiling from well logs. This holds the best prospect for the future of intelligent sand management
2. Whole core sampling, which is limited by core damage
3. Side-wall cores – One-inch plugs in clearly identified pay sections
4. Representative samples via mud logging/sedimentological studies of drilled cuttings from sandstone pay sections

6.5.2 Particle Size Analysis

Knowledge of the grain size and distribution are vital to the design of screen systems slot openings and gravelpack proppant size. The physical particle size analysis can be carried out in three ways:

- Dry sieve analysis using dedicated test sieves
- Wet sieving
- Use of laser particle size analyser (LPA); choice of LPA must consider obscuration level of the LPA.

Examples of commercial LPAs include Malvern and Quantimet Analysers. The advantage of the LPA is the small volume of sample required (less than 2 g). Dry sieve analysis on the other hand may require a minimum of 20 g of sample. Typical formation or particle grain sizes categorised in accordance with the Wenthworth grain sizes and percentiles are given in Figure 6.1.

Typical grain sizes can be represented in inches, mm, microns, or mesh sizes. The mesh size is inversely proportional to the actual grain size. The percentile sizes can be in d5, d10, d20, all the way to d90. The d90 is known as 90th percentile size, and this represents the smallest test sieve size that will capture/trap 90th of the overall particle grain sizes

A typical dry sieve analysis involves the following process:

a. Disaggregate the grains
b. Weigh the sample = x g

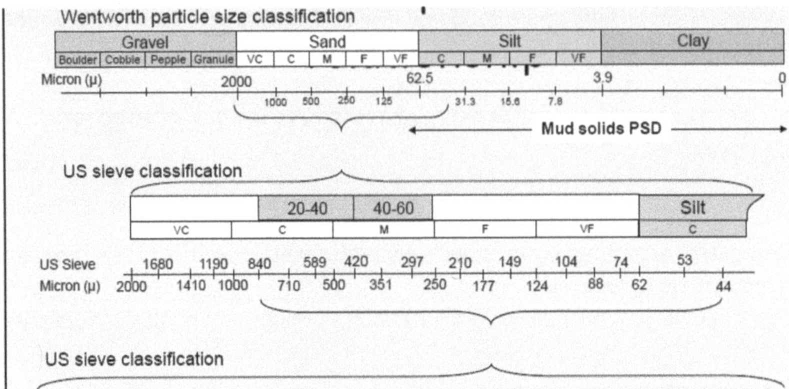

US Mesh	20	25	30	35	40	45	50	60	70	80	100	120	140	170	200	230	270	325
Micron	840	710	589	500	420	351	297	250	210	177	149	124	104	88	74	62	53	44
Inch	0.033	0.028	0.023	0.02	0.017	0.014	0.012	0.01	0.008	0.007	0.006	0.005	0.004	0.004	0.003	0.002	0.002	0.002

FIGURE 6.1 Wentworth Particle Size Classification *(Source: George King[20])*.

FIGURE 6.2 Sieve/Sieve Shaker.

c. Arrange the test sieves in descending order of sieve openings from top to bottom

d. Place the test sieves on a sieve shaker (see Figure 6.2)

e. Pour the weighed particles on to the top sieve

f. Vibrate to standard amplitude with sieve shaker for a given period; usually minimum of 10 min. and above depending on volume

g. Measure each weight captured in each test sieve

Final results are usually given in table form as shown in Table 6.1

TABLE 6.1 Example of Typical Test Sieve Result

Sieve Size, mm	Weight, gm	Weight, %	Cumulative Weight, %
0.425	10	7.193	7.193
0.355	10	7.193	14.386
0.30	15	10.79	25.17
0.25	10	7.19	32.37
0.212	5	3.6	35.96
0.18	3.96	2.85	38.81
0.15	15.07	10.84	49.65
0.125	0	0	49.65
0.09	50	35.963	85.62
0.063	10	7.193	92.81
0.045	10	7..193	100
0.032	0	0	100

From the above data, the Ogive can be plotted as semi-log plot of cumulative % vs sieve size (Figure 6.3). The weight percentage can also be plotted against the sieve size.

LPA vs Dry/Wet Sieve Analysis

The LPA is more versatile in that it can generate the Ogive, weight %, and volumetric concentrations of each particle size in one step. Nevertheless, experience shows that the dry/wet sieve analysis methods provide better representation of particle size. For the LPA the challenge is the obscuration level, which can affect the result outcome.

From the above, the following key textural properties can be computed:

1. Mean size = d50
2. Sorting
3. Uniformity coefficient

For the example case in Figure 6.3, the mean size d50 = 100 μm.

Sorting

This represents the overall grain size distribution of the formation sand. Typical formulas for the analysis are:

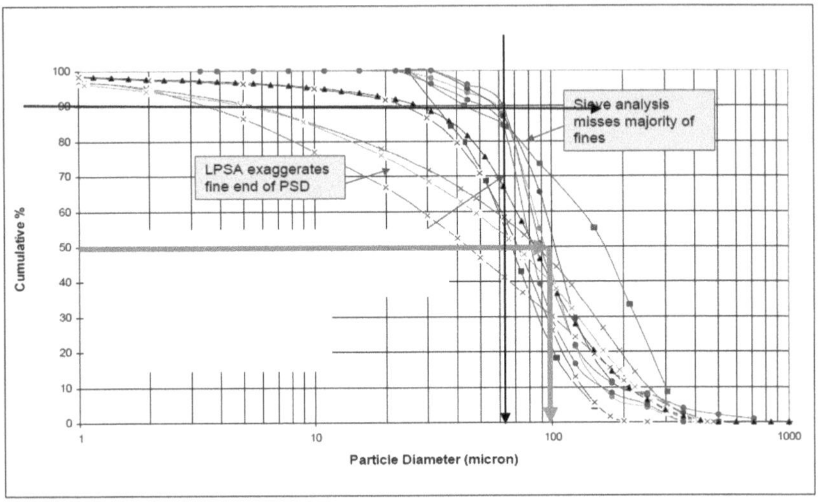

FIGURE 6.3 Example of a Typical Formation Sand Size Distribution Plot, The Ogive.

Sedimentologist's Formula:

$$S = \frac{\phi_{84} - \phi_{16}}{4} + \frac{\phi_{95} - \phi_5}{6.6} \qquad (6.1)$$
$$\phi = -\log_2 d$$

where:

ϕ = Phi unit and (d) = mm

The criteria for sorting classification are:

S < 0.35 - Very well sorted	0.35 < S < 0.5 - Well sorted
0.5 < S < 0.71 - Moderately well sorted	0.71 < S < 1 - Moderately sorted
1 < S < 2 - Poorly sorted	S > 2 - Very poorly sorted

From Figure 6.3, for example:

$\phi84 = 84^{th}$ percentile size of the grain = 3.88
$\phi16 = 16^{th}$ percentile size = 2.18
$\phi95 = 95^{th}$ percentile size = 4.18
$\phi5 = 5^{th}$ percentile size = 1.74

Therefore, S = 1.91, which means the sand is poorly sorted.

Berg's correlation

$$S = \frac{\phi_{90} - \phi_{10}}{2} \qquad (6.2)$$

S < 1 - Well Sorted   S > 1 - Poorly Sorted

where:

$\phi 90 = 3.94$
$\phi 10 = 1.89$

For the example, the case above:

S = 1.025 = Poorly sorted

Uniformity Coefficient

This is defined as:

$$C = \frac{d_{40}}{d_{90}} \tag{6.3}$$

where:

d40 = 40th percentile size
d90 is 90th percentile
For C < 3 = Uniform sand
For C > 5 = Non-uniform sand

6.6 DOWNHOLE SCREEN SYSTEMS

The completion of reservoirs with sanding problems is rather complex, and different methods are available for controlling sand production. An effective sand control method must meet the following key objectives:

- Maximisation of production capacity
- Prevention of fines and load-bearing sand migration
- Minimisation of impairment due to particulate plugging, compression, scale, etc.

Achievement of these key objectives requires the adoption of an integrated approach to the completion design for reservoirs with sanding problems. This should emphasise proper planning of appropriate design criteria and the preparation of pay zones prior to the application of any of the sand control methods.

For screen system completion in particular, this may require a radical change to the overall completion techniques/patterns. Lack of a cohesive approach to screen completions has been the main reason for many of the screen failures reported around the world, especially for horizontal wells. The failures have been largely attributed to the following[1–5]:

1. Screen plugging caused by:
   - High-pressure drop across screens
   - Hot spots of localised production due to localised inflow and annular flow
   - Fine and shale sand
2. Incorrect procedures, materials, or equipment selection including:
   - Installation problems for screens
   - Corrosion due to acid
   - Improper well bore cleanup
   - Ineffective oil-base mud removal
   - Ineffective sand control
   - Inappropriate screen selection
   - Screen erosion especially in gas wells
3. Poor reservoir characterisation especially in the areas of:
   - Reservoir fracturing
   - Grain size textural properties – size distribution, sorting, etc.
   - Sanding up due to water production

Any effective screen completion must adhere to an integrated completion strategy that incorporates not only the effective management/control of solids migration from unconsolidated reservoirs but also accounts for effective management of well bore particulates through:

1. Rational fluids optimisation through effective drilling/completion fluid solids control, choice of fluid type, and recipe
2. Optimum well bore cleanup
3. Optimum screen running techniques
4. Optimum risk assessment and screen coupon testing prior to selection

The key issues facing many of today's engineers once a screen system has been chosen are discussed in the following sections.

Types of Screens

There are now more than nine different screen systems on the market but currently no selection criteria for individual screens. Examples of commercial screens and corresponding vendors are presented in Table 6.2.

The different screen systems are discussed in the following.

6.6.1 Slotted Liners

Slotted liners (SL) are the simplest and cheapest screens. With solid-state laser technology, high-quality slots (0.006") are achievable in all commonly used materials for oil tubulars, including stainless steel. SL are more prone to plugging due to the slot configuration and the smaller inflow

TABLE 6.2 Examples of Commercial Screen Systems

S/N	Screen type	Vendor
1.	Rod-Based & Pipe-Based Wire-wrapped Screen	Johnson Screen Company
2.	Premium Screen	Baker Oil Tools
3.	Excluder Screen	Baker Oil Tools
4.	Stratapac/Stratacoil	Pall Corporation
5.	Conslot Screen	Conslot Corporation
6.	Expandable Screen System (ESS)	Weatherford
7.	ResSHUNT, ResQ	Reslink/Schlumberger
8.	Purolator/PoroPlus	Wesco/Halliburton
9.	Meshrite	Secure

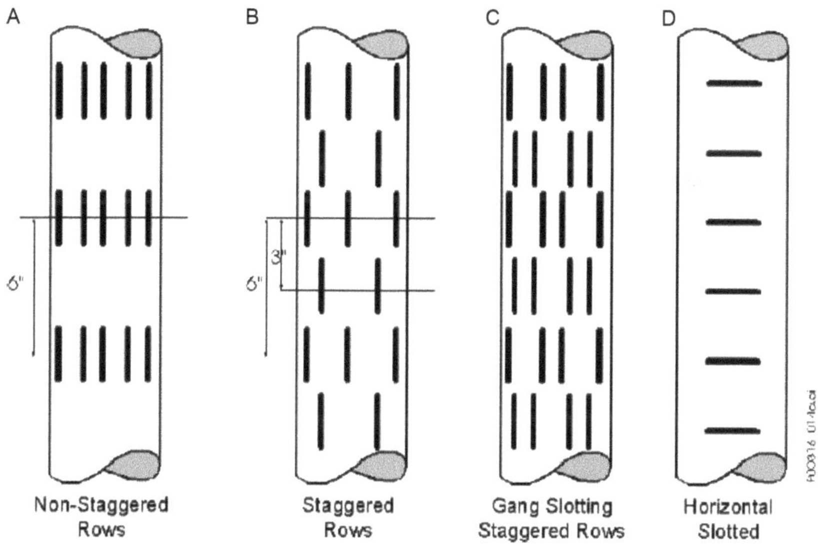

FIGURE 6.4 Slotted Liner Configurations.

area. Slot plugging may be caused by hydrocarbon, scale precipitation, or corrosion. Slotted liners are used predominantly in shallow onshore and low-cost operations where minimizing cost requires some compromise on quality.

"Houston" WWS

FIGURE 6.5 Pipe-Based Wire-Wrapped Screens.

6.6.2 Rod-Based or Pipe-Based Wire-Wrapped Screens

Wire-wrapped screens (WWS) are constructed by wrapping a piece of wire around a perforated or slotted pipe or rod-based system. The wire is spaced to give the required slot width and a self-cleaning slot is obtained by using a "key stone" shaped wire. WWS are more susceptible to erosion than other screens. Small slot sizes lead to reduced inflow area, increased turbulence, and increased plugging risks.

WWS are frequently deployed in medium- to coarse-grained reservoirs. WWS have a much larger flow area than slotted liners, although there are no known cases of production constraints when using slotted liners. WWS is more expensive compared to the slotted liner and can also be used as a back up to slotted liners should the liner not satisfy the sand exclusion requirements. Smaller size WWS have been successfully used inside slotted liners to minimise sand production.

6.6.3 Rod-Based Welded Wire-Wrapped Screen (Con-Slot Screens)

The all-welded or con-slot screens possess more favourable mechanical characteristics than the WWS systems. Screens can be manufactured to 50 μm (0.002") and can be sized to retain formation sand directly. Con-slot screens have a good Outer Diameter (OD)/inner Diameter (ID) ratio. In addition, the flow patterns through the screens are improved through the screen design, which has a very sharp cornered profile (reducing the risk of screen plugging). Material specifications are acceptable for CO_2 service and can be oxidized.

FIGURE 6.6 Rod-based Welded Wire-Wrapped Screens.

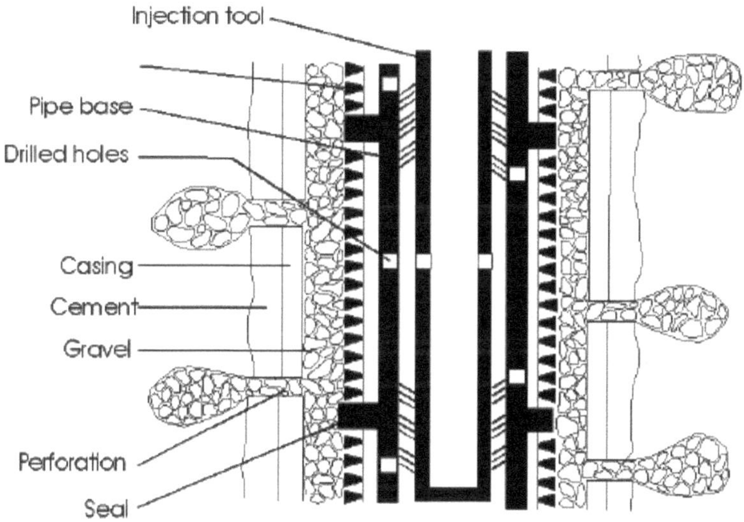

FIGURE 6.7 Selective Zonal Isolation Screen Schematic.

6.6.4 Selective Isolation Screens

This is a variation of the conventional WWS and was invented to enhance the success ratio of remedial treatments through gravelpacked completions in cased hole. The screen is expensive.

6.6.5 Pre-Packed Liners (PPL)

Pre-packed liners are an assembly of concentric liners. PPLs are costlier than slotted lines.

FIGURE 6.8 Premium Pre-packed Screen.

6.6.6 Pre-Packed Screen (PPS)

Pre-packed screens are simply an assembly of concentric screens packed with consolidated proppants. The screens are more flexible when it comes to the design of the sand retention capability. However, they are susceptible to mechanical damage, particularly in high angle and horizontal wells where the screens come into contact with the casing/open hole. Resin bonding the gravel minimises the risk of losing the pack, but this makes the pack brittle and it may crack when subjected to bending. The resins used to bond gravel are not tolerant to stimulation acids. The OD to ID ratio for this form of sand control is generally poor. There are three types of pre-packed screens:

- Single-screen pre-pack: Consists of an inner perforated base pipe surrounded by an inner screen. This is the most commonly used pre-packed screen and the most rugged.
- Dual-screen pre-pack: Consists of an inner perforated base pipe surrounded by screens. This is mostly used in high rate gas wells.
- Thin-body pre-packed screens: Used extensively as replacements for wire-wrapped screens.

6.6.7 Porous Sintered Metal Screens (Sinterpak)

Sinterpaks are formed from sintered stainless steel (or other alloy) bonded under high temperature and pressure and shaped into a cylindrical body. The outer sintered screen is welded to a perforated base pipe. Sintered metal screens are rugged, acid-resistant (including mud acids), and suited to

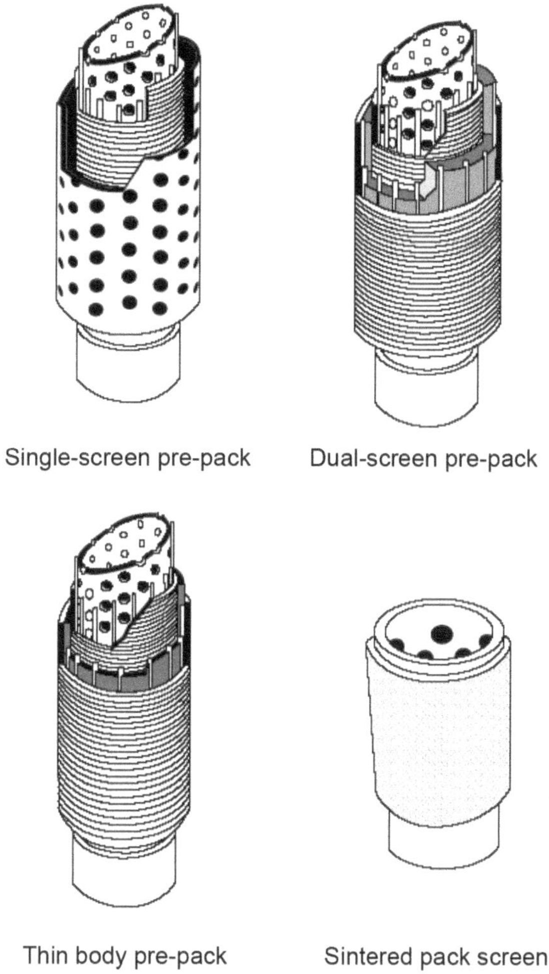

FIGURE 6.9 Different Types of Pre-packed Screens.

doglegged, highly deviated, and horizontal wells. The screens are susceptible to plugging due to the small-pore throat size. Cost is a major issue with this type of sand control, since Sinterpaks cost twice as much as wire-wrapped screens.

6.6.8 Porous Metal Membrane Screens (Stratapac and Stratacoil)

A comparatively new technology to conventional screen design is Stratapac screens. The screens consist of a metallic screen made of a composite of thin

FIGURE 6.10 Porous Sintered Metal Screens.

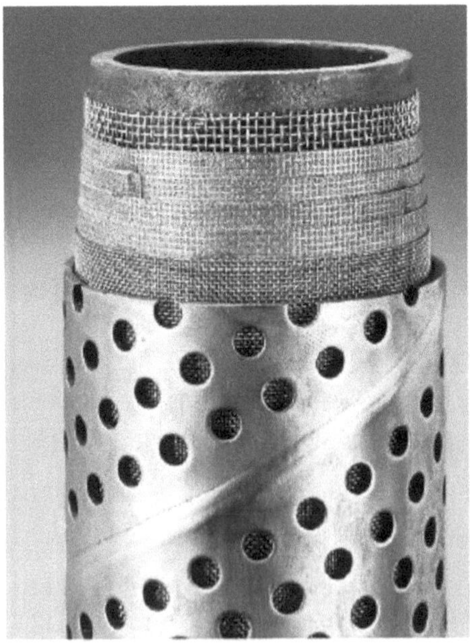

FIGURE 6.11 Stratapac *(Source: http://www.weatherford.com[21])*.

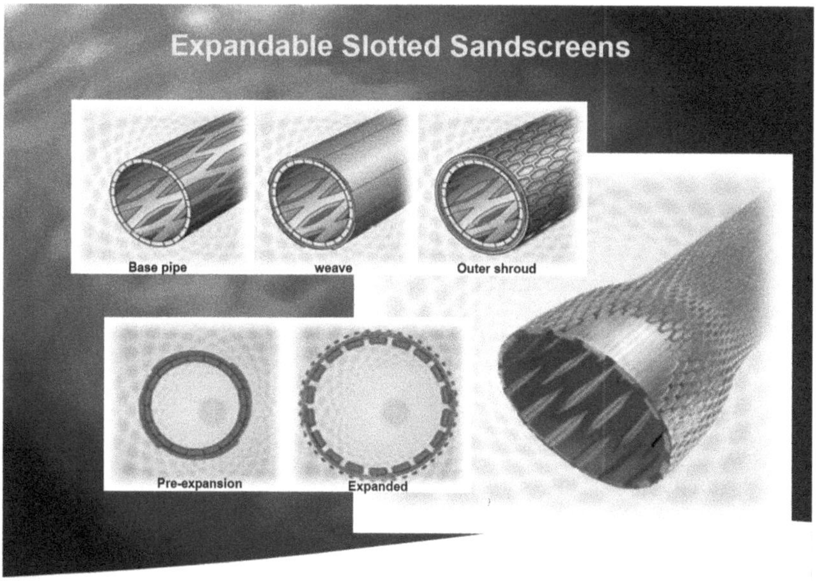

FIGURE 6.12 Expandable Sand Screen *(Source: http://www.weatherford.com[20])*.

stainless-steel layers of porous metal membranes sandwiched between a perforated API base pipe on the inside and a protective cage (shroud) as the outer jacket. The porous metal membrane is highly flexible and ductile, permitting extreme deformation without loss of integrity. Coiled and through tubing screens (Stratacoil) are also available that have the critical OD requirements to allow passage through small tubulars.

Stratapac screens are commonly used for low-cost remedial sand control. The screens have a compact design with low ID/OD ratios and are claimed to have superior inflow characteristics over pre-packed screens, have a greater resistance to damage, and have high tensile strength.

Stratapac screens have also been used for primary sand control, usually in conjunction with sized gravel (i.e., as a more rugged form of retention screen). An attractive feature of this form of screen is their mechanical robustness with respect to other screens. They keep their retention capability when bent or crushed and are resistant to acids.

6.6.9 Expandable Sand Screens

This is the most commonly used screen. The expandable sand screen (ESS) has a slotted or perforated base pipe – for borehole support and screens – for sand control (in various mesh sizes) and an outer protective shroud – which

FIGURE 6.13 ResSHUNT by Reslink *(Source: http://www.slb.com[22]).*

protects the screen during deployment and service life. The main attributes of ESS are:

- Larger inflow area
- Easy to install
- Good borehole support

Field results using ESS to date has revealed the following advantages:

- Low skin values in wells
- Does not allow flow outside the screen leading to a reduction in screen erosion
- Bigger internal diameter allows easy access for well intervention and any problematic zones can be isolated independently with clads
- High resistance to erosion, corrosion, screen plugging, and collapse

Table 6.3 gives the advantages and disadvantages of some current screen designs.

TABLE 6.3 Active Sand Control Methods – Screen Systems

Screen Type	Advantages	Disadvantages
Conslot	– Low plugging tendency – Easy to clean up – Gauge can be tailored to formation – Suitable for high water cut wells	– Average damage tolerance – Fine production may lead to topside or artificial lift component problems
WWS	– Low plugging tendency – Easy to clean up – Very suitable for uniform, large- grained formations – Corrosion resistance is high – Low cost option	– Average damage tolerance – Fines production may lead to topsides or artificial lift component problems – Limited erosion resistance especially when oversized – Quality control is critical with respect to slot width
Excluder	– High solid resistance – Excellent clean up of plugging if drilling fluids have no drilled solids – Erosion-resistant because of vector shroud – High corrosion resistant	– Fine production may lead to topsides or artificial lift component problems – Expensive
Sinterpack	– Acceptable clean-up if drilling fluids contain no drill solids – Low ID/OD ratio with respect to PPS	– Very low solids tolerance – Low erosion resistance of the pack – Low damage tolerance
Stratapac	– Damage tolerant – Acceptable clean-up if drilling fluid contains no drill solids – Corrosion resistance high – Zero sand production – Low ID/OD ratio with respect to pre-packs	– Very low solids tolerance – Expensive
Stratacoil	– Through tubing application – Large ID for given OD sizing – Damage tolerant – Suitable for remedial sand control – Low ID/OD ratio with respect to pre-packs	– Not suitable for formations producing at high fine and clay rate (low solid tolerance)

6.6.10 Special Screen Systems

ResSHUNT

ResSHUNT™ is a new generation sand screen for gravelpack and frac and pack applications employing shunt technology and is manufactured by RESLINK. Features include:

- High flow area compared to existing solution (5 ½" base pipe/7 ½" OD or 6 5/8" base pipe/8.2" OD)
- Screens are handled and installed as casing
- Base pipe is conventional casing -long joints (+40 ft)
- All forces onto screen jacket are taken up by base pipe
- No hang up of shunt tubes during installation
- Nozzles can be changed on location
- One connection to make up on the rig floor
- No jumper connectors to make
- Significantly reduced make-up and installation time
- Four or more 1" or ¾"shunt tubes
- Based on Reslink's robust LineSlot™ screen design
- Shunt tubes are protected between base pipe and wire wrapping
- Accurate and consistent slot opening for superior sand control

Benefits include:

- Facilitate complete placement of sand across the entire interval
- Reduced rig time – joints are made up like casing
- Robustness and smooth external surface ensure risk-free placement in open-hole and cased-hole sections
- High ID for efficient hydrocarbon production

ResQ

ResQ™ is a pre-pack screen utilizing a knitted stainless-steel wire jacket as a particle discriminating medium. It is an energy-absorbing sand screen. The

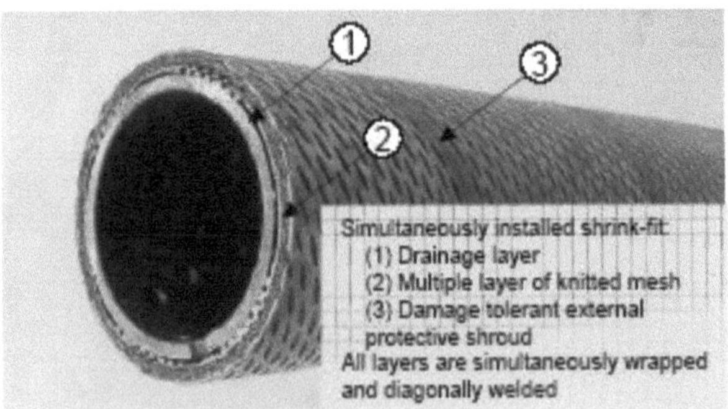

FIGURE 6.14 ResQ by Reslink[22]. UltiPac by Schlumberger.

objective is to provide a sand screen with enhanced erosion resistance. The interlocked and looped structure of knitted wire makes it robust and large loops (2–3 mm or 0.08–0.12") minimise risk for trapping fines. The knitted wire medium can be manufactured from a variety of steel wire thickness, material types, number of wire strands in the knit, etc. The density and thereby porosity can be modified by the degree of compaction the knitted wire medium is exposed to in the manufacturing process.

6.7 SCREEN DESIGN CRITERIA

Screen design criteria include but are not limited to screen slot gauge (for the outer wire-wrap, pre-pack thickness (for pre-packed screens), pre-pack proppant size and type (for pre-packed screens), base-pipe perforation density, screen jacket diameter, and how best to run the screen with or without displacement of the screen to drilling fluid. Other key issues are analysis of formation grain size textural properties (details of the formation sampling and textural analysis are presented in Section 6.5) and management of well bore fluids including the selection of loss control materials.

6.7.1 Screen Slot Gauge

Current industry practice for the outer screen slot gauge selection is generally based on the 10th percentile size of the candidate sand.

- Coberly Criterion $= 2 - 3\ d_{10}$ Formation Sand

The modified Coberly rule for slot gauge selection is defined as:

- Gauge $= 2 \times d_{10}$ formation sand
- Gauge $= d_{10}$ formation sand
- Gauge $= d_{50} \times$ formation sand (Conslot)

The criteria assume minimum sand production before bridging takes place.

For sand production levels indicated, the conditions are:

High: $> 10\%$
Medium: $<10\%$
Low: $<1\%$

6.7.2 Proppant Size and Pre-Pack Thickness

1. Proppant Size
   With current industry practice, proppant sizes are still selected based on the Saucier or modified Saucier rule defined as:
   - d_{50} proppant $= 5-6 \times d_{50}$ Formation sand
     or
   - d_{50} proppant $= 4-8 \times d_{50}$ Formation sand

For synthetic proppants, it has been suggested that larger grain sizes than predicted by Saucier rule will be effective.

2. Pre-pack Thickness
There are no established criteria for the pre-pack thickness. Depending on the manufacturer, they range from 0.23" for low-profile pre-pack (Halliburton and Johnson screens) to sometimes up to about 0.6" for Ultra-pack (Johnson screen). Screen manufacturers do provide general guide to the selection of pre-pack thicknesses. The best approach is to carry out a full-scale bench test for the effectiveness of different pre-pack thicknesses. Overall, recent studies on the analysis of the performance of gravelpacks have shown that the Saucier rule is too general and gravel sizes may be far smaller or larger than what is proposed by the Saucier rule depending on operating conditions.
3. Base pipe perforation density
Different vendors have different base-pipe perforations. The ranges are:
A. 2-7/8" Base Pipe
Halliburton = 102
Johnson = 108
B. 3-1/2" OD
Halliburton = 126
Johnson = 108
For selection of base pipe perforation, the recommended underlying factor should be proper inflow area and the friction factor.
4. Screen Jacket OD
This depends largely on the type of screen and vendor's specification in relation to hole size. Standard practice adopts the conventional API standard for casing size in open hole.
Screen Dimensions: The size of a screen is denoted by the nominal outer diameter (OD) of the base-pipe. The optimum dimensions of a screen for a given hole size is a compromise between the following factors:

Typical Screen Picks	mesh	inch	millimeters	microns
	8	0.093	2.38	2380
	10	0.078	2	2000
32 gauge →	12	0.066	1.68	1680
	14	0.056	1.41	1410
	16	0.047	1.19	1190
	18	0.0394	1	1000
24 gauge →	20	0.0331	0.84	840
	25	0.0278	0.71	710
18 gauge →	30	0.0234	0.59	590
	35	0.0197	0.5	500
12 gauge →	40	0.0165	0.42	420
	45	0.0139	0.35	350
	50	0.0117	0.3	300
8 gauge for ceramic / 6 gauge for sand →	60	0.0098	0.25	250
→	70	0.0083	0.21	210

FIGURE 6.15 Typical Screen Slots Selection (*Source: George King*[20]). MeshSlot by Schlumberger.

- Maximum inner diameter (ID) to reduce to a minimum the frictional pressure drops and maximise tool accessibility.
- Minimise clearance between screen OD and sand face to reduce to a minimum movement of sand and fluids behind the screen.
- Sufficient clearance between the screen and the sand face to cater for optimal installation.

Screen Material Selection: Screen material specifications should be assessed with a priori knowledge of the anticipated production conditions. The primary concern for a screen is to ensure filter integrity over the field lifecycle. Screens made of special alloys are obviously more expensive and quality control specifications more stringent. Screens are generally made of AISI 304L or AISI 316L stainless steel for the jacket (AISI - American Iron and Steel Institute). In most instances, the base pipe is made of carbon steel. Both AISI specifications are good for CO_2 service in producing wells and up to at least 120 °C. These alloys are, however, susceptible to pitting, crevice, and stress corrosion cracking (SCC) in the presence of chlorides and oxygen (AISI 316L has a higher corrosion resistance than AISI 304L). To reduce the risk of corrosion, the pH of the workover/drilling brine should be kept above 9 and when extended exposure is unavoidable, an oxygen scavenger should be used. Alloys such as Sanicro 28 or Incoloy 825 are adequate for higher H_2S levels (up to 5 bar partial pressure), with or without CO_2 and for the same temperature range.

Required Inflow Area: The largest number of uniformly distributed slots or pores should be targeted, without affecting the mechanical integrity of the liner, keeping in mind the mechanical loading during installation and operation. However, for slotted liners, there is a clear economic incentive to minimiseing the number of slots to be cut, as costs are proportional to the number of slots. A large inflow area is beneficial in maintaining a low velocity in the slot area, thereby reducing erosion.

6.7.3 Screen Selection Criteria

There are no known standard industry criteria for screen selection. Overall, rigorous tests must be carried out in order to select the optimum criteria for individual cases.

In general, screen selection criteria should be based both on:

- Benchmark plugging test
- Simple stochastic analysis of how the screen would respond to inflow from the reservoir

The plugging test should focus on the following operating parameters to benchmark against the impairment of the screen systems by solids:

1. Flow rate and pressure
2. Solids textural properties:
   - Size
   - Size distribution (sorting)
   - Shape

Downhole screens are designed to allow the majority of formation particles to 'bridge' across the openings, yet offer maximum fluid flow area. Smaller formation particles are then retained behind the larger "bridged" particles. A wide variety of screens are available, and these are discussed in this section. The main advantages, disadvantages, applicability, and design criteria of screens are as follows.

Advantages

- Best results seen in horizontal wells
- The installation is cheap and simple

Disadvantages

- Manufacturing smaller screen openings is somewhat difficult
- Finer sand requires smaller screen openings leading to more turbulence at the openings
- During installations and production, screens have a higher probability of being plugged
- Low to mid erosional velocities are capable of washing out screens
- High in-situ stresses can result in mechanical failure

Applicability

- Can be used in any wells; even in highly deviated/horizontal wells

6.8 GRAVELPACK COMPLETION

Gravelpacks provide sand control by the filtering capability of gravel that is kept in place by a sand screen. Optimally sized gravel is placed against the formation and positioned in place by a screen that is typically the wire-wrapped type (Figure 6.5). This gravel is sized to hold back the formation sand. The gravelpack can be placed either in perforated casing (IGP – internal gravelpack) or open hole (EGP – external gravelpack). The open hole is generally under-reamed prior to placing the gravel. This can also be done after milling a section of the casing away (milled casing under-reamed gravelpack – MCUGP).

The gravelpacking process generally involves effective well bore cleanup followed by the placement of the gravelpack screen in the zone opposite the pay section. This is followed by the pumping of the appropriate

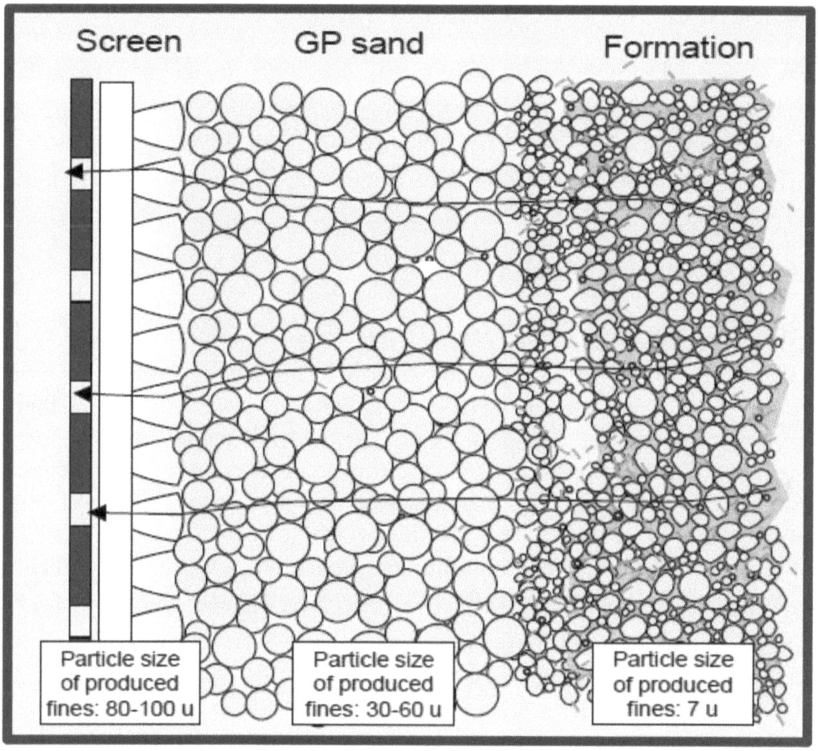

FIGURE 6.16 Typical Open-Hole Gravelpack.

concentration of gravel slurry into the annulus between the screen and the sand face or between the screen and the casing and into the perforations until screen out. The gravel settles in the annulus while the carrier fluid leaks through the screen into the wash pipe and is returned to the surface (Figure 6.5).

Internal Gravelpack

IGPs are applied in wells that are cased and perforated, and are the most widely used method with a high degree of reliability and flexibility. The placement of gravel is however more complex than for EGP largely resulting from the smaller annular space, and because the gravel needs to be squeezed through the perforations. Work-overs are difficult (especially for multi zone completions) as screen, packers, and gravel usually need to be removed prior to a repair or re-completion. IGPs are applicable in a wide range of formation types (e.g., sand/shale sequences, sand with high clay or silt content). Low productivity is often observed although theoretically the productivity index (PI) should be that of perforated completions.

6.8.1 Design and Installation of Gravelpacks

The design and installation of gravelpacks depend on the following key factors:

1. Well preparation
2. Formation sand characterisation
3. Gravel or pack sand selection
4. Gravel placement design
5. Post gravelpack evaluation

Well Preparation

It is important to utilise low or solids-free reservoir drilling fluids for the pay section. Such fluids should include sized salts or properly sized calcium-carbonate loss-control materials. Well bore cleanup including removal of filter cake via liftoff and use of cake solvent breaker prior to gravelpack installation is very important. Effective well bore cleanup to remove well bore debris with low or solids-free, low Equivalent Circulation Density (ECD) cleanup fluids essential to minimising gravelpack plugging and impairment.

For cased-hole completions the following are essential:

- Cake removal before setting casing
- Casing cleaning: Scrapping/solvent
- Deep perforation penetration
- Perforation cleaning
  - Underbalanced perforating/backsurging
  - Perforation washing: Recommended for highly unconsolidated sand
- Perforation size/Density: 1/2"/8-12spf
- 90° phasing/'Ten-to-Two' or 'Quarter-to-Three' configuration

Formation Sand Characterization

To effectively characterize a formation for the purposes of gravelpacking, the grain size distribution must be known as originally described in Section 6.8. The grain size distribution is determined by sieving the sand grains over a series of sieves with declining mesh size. The results of the sieve analysis are plotted as a cumulative weight percentage retained at the sieve versus the mesh opening (grain size) on a semi-logarithmic scale. From this plot, the following parameters are determined:

- Median size D_{50}: This is the particle size corresponding to the 50th percentile size of the size analysis
- Sand sorting
- Uniformity coefficient: Calculation as D_{40}/D_{90}
- Shape of the sand

Gravel Selection

The gravel selection criteria includes:

- Overall sand size distribution, sorting, and shape
- Gravel type, shape and structure
- Bridging efficiency of gravelpack at prevailing operating condition and well bore configuration
- Prevailing pore-blocking mechanism
- Short-term vs long-term performance

A few factors need to be evaluated in selecting the optimum gravel. These factors are include gravel sizing, quality, and permeability

Gravel Sizing

(a) Gravel sizing: Most widely used criterion for gravel sizing is the Saucier criterion, defined here as $d_{50}(\text{Gravel}) = 5\text{-}6 \times d_{50}(\text{Sand})$

(b) For poorly sorted sand, gas wells, open-hole high producers, cased-hole gravelpacks:
Gravel = 3-4 × d_{50}(Sand)

(c) For low producers, uniformly sorted sand in open-hole completions:
Gravel = Saucier rule

(d) For synthetic gravels:
Gravel = 6-7 × d_{50}(Sand)

TABLE 6.4 Typical Commercial Gravels

Gravel Size, μm	US Mesh Size	Av. Size, μm
150 × 425	40/100	300
200 × 425	40/70	330
250 × 425	40/60	350
425 × 825	20/40	630
575 × 1175	16/30	880
825 × 1650	12/20	1260
975 × 1650	12/18	1340
825 × 1975	10/20	1410
1175 × 1975	10/16	1590
1650 × 2350	8/12	2020

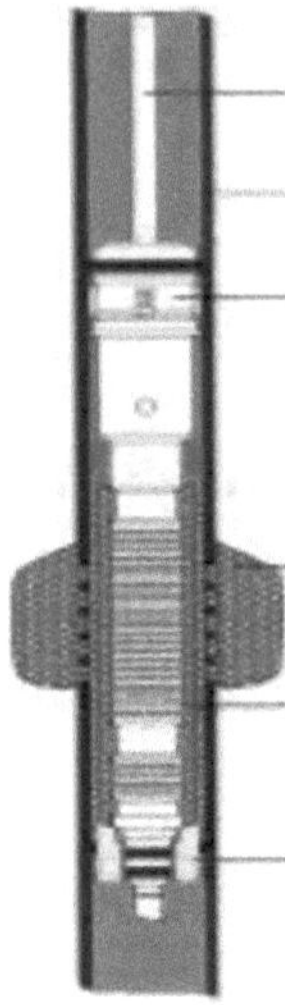

FIGURE 6.17 Completion Configuration for Gravelpack.

Synthetic proppants are popular for pre-packing perforations, and low-density gravel systems (LDGS) are also popular for gravelpacking long horizontal and ER wells.

Gravel Quality

Typical characteristics of good quality gravel are high roundness and sphericity, high acid resistance, high crushing resistance, low percentage of multi-crystalline grains, and low silt and clay content. This will inhibit the formation of filter cake at the perforation wall.

Gravel Permeability

Permeability of a loosely packed gravel bed is higher but sand retention capability is reduced. Particularly in IGPs, this results in gravel/sand mixing and invasion and permeability reduction.

Gravelpack Screen Slot

This is designed against the retention of the gravelpack. The gravelpack screen slot size = 90th percentile (d90) of the pack sand size distribution.

Gravel Placement Design

The key factors to take into account in terms of placement design are:

1. Length of interval
2. Gravel concentration

3. Carrier fluid type and properties
4. Completion technique – Open-hole or cased-hole completion
5. Wash-pipe size. This is equal to $0.8 \times$ Screen ID
6. Placement technique and slurry pack versus squeeze pack

The placement techniques include:

- For Open-hole Completion: The crossover tool/shunt packing or slurry-packing techniques are the most popular. For highly deviated and horizontal wells the alpha and beta wave slurrypack technique combined with optimum wash-pipe size is the most popular.
- For cased-hole completion, the two-stage squeeze packing of perforations followed by conventional annular packing is common

Gravelpack Problems

- Evidence of fines production in large quantities
- Evidence of high positive skin
  - Poor gravel sizing
  - Fines/sand invasion

6.9 CHEMICAL CONSOLIDATION

Sand CONsolidation (SCON) is a sand control method whereby fluids containing a cementing agent are injected into the formation to provide a bond between sand grains after curing. As the sand grains are coated and bonded together by the cementing compound, some permeability will be lost following the treatment. The successful treatment must provide the additional strength required while preserving as much formation permeability as possible. Various types of chemical consolidation systems are available, e.g., EPOSAND and WELLFIX.

Advantages

- Sand consolidation can simplify multiple completions (no hardware obstructing the well bore) and reduce workover costs
- Less apparent impairment than gravelpacks
- Rigless installation possible (through tubing or coiled tubing treatment) and hence inexpensive

Disadvantages

- Optimal placement of the treatment fluids is critical for success. The treated interval is constrained to some 3 to 4 m, although this restriction may be overcome with the use of new tools/techniques such as a pinpoint packer.

Applicability

- Consolidation methods constrained to relatively clean formation sand (i.e., Net-to-Gross >= 75%)
- Best applied where formation permeabilities are greater than 500 mD. Low permeability sand may be impaired by the consolidation treatment.

6.9.1 Chemical Consolidation Chemical Types and Procedures

- Treat formation in immediate vicinity of well bore to bond sand grains
- Formation must be treated through all perforations
- Consolidated sand mass must remain permeable to well fluids
- Consolidation should remain constant over time
- Two principal types of treatment
  - Epoxy resin (three-stage treatment) – isopropyl alcohol pre-flush, then resin is pumped followed by viscous oil to displace resin from the pore space). Limitations – only 20 ft at a time, temperature maximum of 100 °C, max clay content 20%.
  - Furan, phenolic resins, and alkoxysilane – have higher temperature range than epoxy but consolidation may experience brittle failure. They are difficult chemicals to handle safely.

Design/Selection Criteria

As with other sand control methods, a few factors need to be considered before embarking on a sand consolidation job. These factors are:

- Temperature: Because of the organic polymer nature of the sand consolidation system, many of its critical properties are temperature dependent, before, during, and after curing. For correct placement, the system

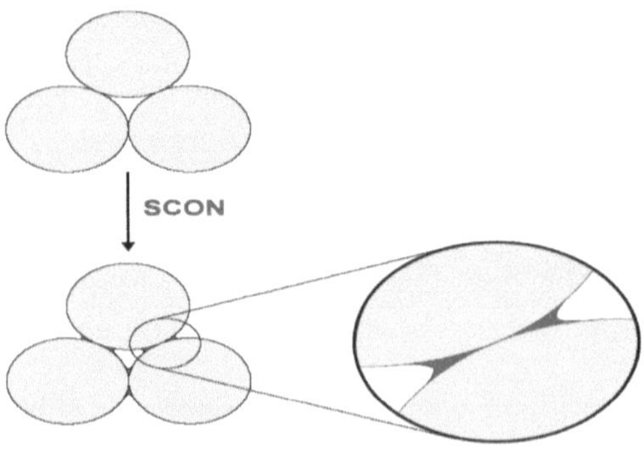

FIGURE 6.18 Chemical Consolidation.

needs to be pumped through the conduits and into the formation before its viscosity increases to such an extent that it can no longer be injected. In principle, the maximum temperature at which this can be safely done poses an upper limit to the bottom-hole temperature. In practice, it is feasible in a number of cases to cool the well down by bull-heading brine down the tubing and into the formation. Finally, the tools necessary to ensure correct placement of the sand consolidation system (such as inflatable packers) may also pose an upper limit to the bottom-hole temperature. This limit is dependent on the design and is available from tool suppliers.

- Type of formation: In general, sand consolidation can be applied to sandstone and loose sand formations that may contain various contaminants, such as clays. However, clay-contaminated zones often have low injectivities that are too low to make sand consolidation practical. In such a case, an acid job is an option to increase injectivity.
- Fracture propagation pressure: Fractures are high-conductivity avenues. These high-conductivity avenues may lead to poor distribution of the resin around the well bore including development of untreated low permeability streaks. Under this scenario, the sand consolidation system can contribute little to sand control.
- Treatment interval: There is a lower limit for the treatment interval caused by the difficulties in correctly placing small volumes of fluid, the accuracy of the depth correlation, etc. However, the actual restriction depends on the tools used. On the other hand, there is no proven upper limit for the treatment interval. But to improve control of the placement, the treatment interval is generally restricted. However, the restriction is generally a qualitative one.
- Formation fluids: As the sand consolidation enters the formation, it encounters the fluids that previously occupied the pore spaces. These formation fluids can interfere with the micro-placement of the sand consolidation system. For instance, gas could prevent the sand consolidation system from reaching all the appropriate crevices, simply by blocking access. Crude oil, on the other hand, can interfere with the adhesion of organic adhesives, leading to a reduction in mechanical strength. The effects of crude oil can be reduced to a minimum by flushing the crude oil from the near-well-bore area of the formation with a suitable pre-flush, e.g., diesel. Finally, brine has an adverse effect on the current sand consolidation systems because it competes with the system in the micro-placement phase. Most sand consolidation systems exploit the fact that their interaction with silicate minerals is much stronger than that of aliphatic mineral oil. However, because of its high polarity and its capacity to engage in hydrogen bridge formation, water bonds even stronger to most silicate type minerals than the sand consolidation system. Water therefore interferes with the micro placement of the resin. This causes

mechanical strength loss and possibly impairment. As a result, the connate brine that is generally present in the formation should be completely removed.

6.10 SAND CONTROL METHOD SELECTION

The key factors taken into consideration in the selection of a sand control technique are:

1. Design complexity
2. Installation complexity
3. Mechanical robustness
4. Well productivity
5. Intervention philosophy
6. Reservoir management requirement
7. Expected reliability
8. Sand quality
9. Multiple intervals
10. Type of well (injector or producer)
11. Expertise available
12. Sanding risk
13. Plugging risk
14. Erosion risk
15. Sand production history

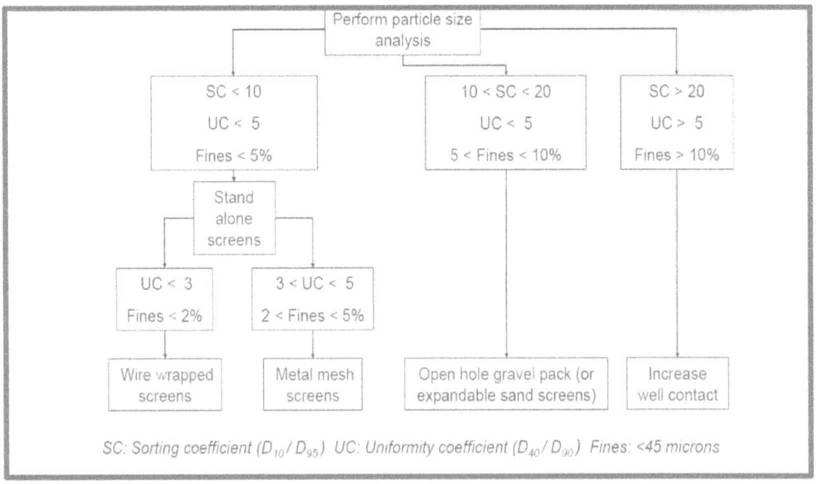

FIGURE 6.19 Rule of Thumb Selection Criteria *(Source: SPE 39437[23])*.

16. Sand tolerance level
17. HSE consideration
18. Single or comingled production

It is very useful to develop appropriate ranking criteria for optimum selection. A general rule of thumb guide is given in Figure 6.19.

Chapter 7

Multiphase Solids Transport

Chapter Outline

7.1 INTRODUCTION

Petroleum as produced from a reservoir rock is a complex mixture of hundreds of different hydrocarbons all with different densities and other physical characteristics. A typical well stream is a high velocity, turbulent, constantly expanding mixture of gases and hydrocarbon liquids, intimately mixed with water vapour, free water, solids, and other contaminants. As it flows from the hot, high-pressure petroleum reservoir, the well stream undergoes continuous pressure and temperature reduction. Gases evolve from liquids, water vapour condensation, and some of the well stream changes in character from liquid to free gas. The gas carries liquid mist droplets and the liquid carries gas bubbles.

The transportation of the hydrocarbon fluids up the well and through the wellheads to the production manifold/process facilities and the pipeline transporting medium (Fig. 7.1) is accompanied by this phenomena change whether

Developments in Petroleum Science, Vol. 63. http://dx.doi.org/10.1016/B978-0-444-62637-0.00007-5

FIGURE 7.1 Long Transport Pipelines.

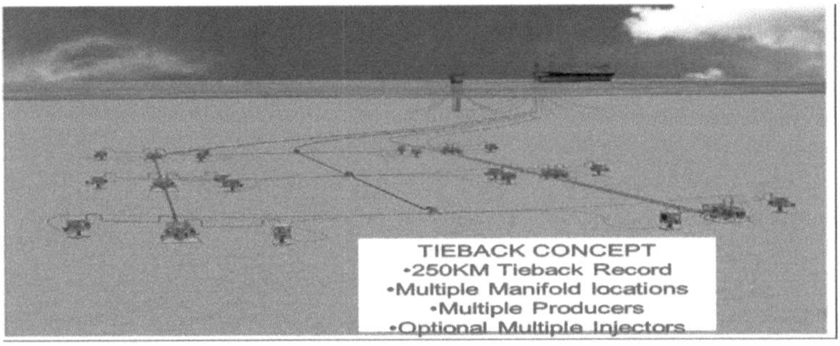

FIGURE 7.2 Example Illustration of Subsea Tiebacks *(Source: http://www.intelligent-flow.com[7]).*

offshore or onshore. Doping of the pipelines with appropriate inhibitors – scale, hydrate inhibitors, for example – and friction reducers adds to the complexity of the multiphase fluids.

For deepwater environments, subsea development with long tiebacks to the central production facility is becoming the most economic, cost-effective method for deepwater development (Figures 7.2 and 7.3). In such deepwater environments multiphase fluid production accompanied by sand production is inevitable. Subsea developments/tiebacks are at high risk averse to sand/multiphase production for effective flow assurance.

In ultra-deepwater and mature fields, multiphase fluid inflow with entrained sand production is inevitable. Sand transport in a multiphase environment is a challenge because of transient solids flow pattern changes and associated huge pressure drops. Solids are transported in various forms – suspension, where velocity is high enough, rolling, and saltation (see Figure 7.4). The key flow assurance issue is that the pipe must be free of sand beds. The sand will settle to form beds and may block the pipeline if the fluid velocity is below the minimum transport velocity required for rolling or saltation. The economic implication of any pipeline blockage is huge given the number of producing wells that may be tied to it. Therefore, understanding the mechanism of sand transport in multiphase flow in pipes, taking into account the

FIGURE 7.3 Subsea Tieback Showing Solids Settling – A Flow Assurance Issue (*Source: Bello*[24]).

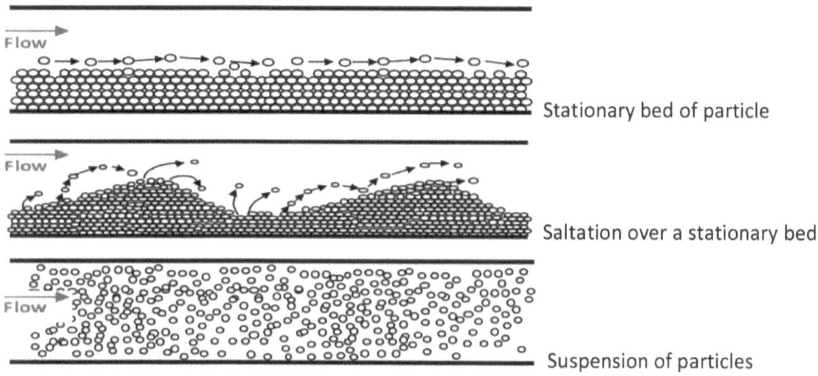

FIGURE 7.4 Sand flow pattern formation in pipe.

continuous changes in flow patterns, has a direct impact on the estimation, design, and detailed analysis of subsea tiebacks.

Generally, different flow patterns are formed whenever there is flow of multiphase fluids in pipelines/tiebacks. The formation of a particular pattern is dependent on flow rates, fluid properties, pipe size, and pressure-drop profiles. Investigating the impact of flow patterns on sand transport in multiphase fluids is critical to sand transport mechanisms. Effective sand transport will require tracking the flow pattern changes as multiphase flow through the pipe.

The accuracy of solid transport models depends on how well the hydrodynamics is described and predicted with sufficient reliability. To establish an accurate predictive model(s) for minimum transport velocity in a multiphase production, it is necessary to understand the mechanism of interaction of fluid flow and sand particle movements. The particles are transported differently by different flow patterns because they are subjected to different driving forces, including lift and drag. The lift and drag coefficients are semi-empirical and demand sufficient experimental and theoretical investigations.

7.2 MULTIPHASE FLOW PATTERNS

Simultaneous passage of gas and liquid in a pipe often results in a variety of flow patterns. Two-phase flow or three-phase flow is simultaneous flow of discrete phases (gas or multiple fluids). These phases are commonly encountered in the petroleum or allied industry. For subsea deepwater developments, multiphase production is inevitable, especially in the tiebacks.

In production tubings and long tiebacks, multiphase fluid is be made up of gas, oil, condensed water, formation water, chemical inhibitors (hydrate and scale inhibitors, anti-agglomerates, etc.), and of course, produced solids. Multiphase fluid flow is transient and accompanied by pattern transition, which is dependent on flow rates, fluid properties, pipe size, well trajectory, and corresponding pressure drop. The concept of flow patterns in pipes introduces new challenges in the understanding of multiphase fluids principally because of the form in which the fluids exist in pipes. The well bore may be horizontal, near horizontal, vertical, or a combination of all depending on the trajectory. For long tiebacks, the route is through different topographies and therefore different pipe angles.

For gas-oil-water-inhibitor phases, the flow patterns can be grouped into four main classes where each class can be subdivided into sub-classes for detailed description as follows:

- Stratified flow (subclasses: stratified smooth, stratified wavy)
- Intermittent flow (subclasses: elongated bubble, slug, churn)
- Annular flow (subclass: wispy annular)
- Bubble flow (subclasses: bubbly, dispersed bubble)

For the purpose of this text, the patterns are grouped as:

- Dispersed bubble
- Annular flow (for low aqueous-gas ration (AGR) in gas producers with formation water, condensed water, hydrate inhibitor, etc.)
- Mist flow
- Slug flow
- Plug flow (in vertical pipe sections)
- Stratified flow (in highly deviated and horizontal pipe sections)

A definition of the multiphase flow patterns is given in Table 7.1. Figure 7.5 illustrates one or more multiphase flow patterns that may exist in a producing well completion string. For a well in production, there is flow pattern transition as the fluid flows up the production tubing accompanied by pressure drop and temperature changes.

Figures 7.6 and 7.7 show general multiphase phase flow patterns for vertical and horizontal pipes depending on the pipeline topography. For vertical pipe multiphase fluid flow the dominating patterns in descending order based on decreasing pressure are:

TABLE 7.1 Definition of Multiphase Flow Patterns

Flow Patterns	Characteristics	Conditions of occurrence
Fluid flow modes		
Annular dispersed flow	The gas flows along the centre of the tube or partially as droplets in the central core. The liquid travels partly in the form of an annulus at the wall.	This occurs at very high gas velocity and low liquid velocity.
Dispersed bubble flow	The gas phase is distributed as discrete bubbles in an axially continuous liquid phase. Increased liquid flow rate prevents bubble accumulations and are dispersed more uniformly in the liquid phase.	This occurs at a very high flow rate. For very high liquid velocities and low gas/liquid ratios, the dispersed bubble flow pattern will prevail.
Slug flow	Slugs of liquid are separated by coalesced gas bubbles. The intermittent pattern is evidenced when fluids are subdivided into slugs and elongated bubble patterns.	For intermediate liquid velocities, rolling waves of liquids will be formed. The rolling waves increase to the point of forming a slug flow, sometimes refer to as plug flow.
Churn flow	This is similar to slug flow pattern but highly disordered in which the vertical motion of the liquid is oscillatory. In this case, the continuity of the liquid in the slug region is destroyed by a high gas concentration.	The liquid and gas rates are intermediate between the annular flow and slug flow for churn flow to occur. Further increase in flow velocity makes the pattern unstable.
Annular dispersed flow (ADF)	The liquid travels partly as a continuous film around the perimeter of the pipe and partly as a small droplets distributed in the gas phase.	This occurs at very high gas velocity and low liquid velocity.
Stratified (wavy) flow (SWF)	This is characterised by separation of fluids into different layers, with lighter fluids flowing above the heavier fluids.	For low flow rates of liquid and gas, a smooth or wavy stratified flow will occur. The interface may be smooth or wavy; hence the term wavy stratified flow.

Continued

TABLE 7.1 Definition of Multiphase Flow Patterns—Cont'd

Flow Patterns	Characteristics	Conditions of occurrence
Slug (intermittent) flow	Slugs of liquid are separated by coalesced gas bubbles. The intermittent pattern is evidenced when fluids are subdivided into slugs and elongated bubble patterns.	For intermediate liquid velocities, rolling waves of liquids will be formed. The rolling waves increase to the point of forming a slug flow, sometimes refer to as plug flow.
Dispersed bubble flow	The gas phase is distributed as discrete bubbles in an axially continuous liquid phase. Increased liquid flow rate prevents bubble accumulations and are dispersed more uniformly in the liquid phase.	This occurs at a very high flow rate. For very high liquid velocities and low gas/liquid ratios, the dispersed bubble flow pattern will prevail.

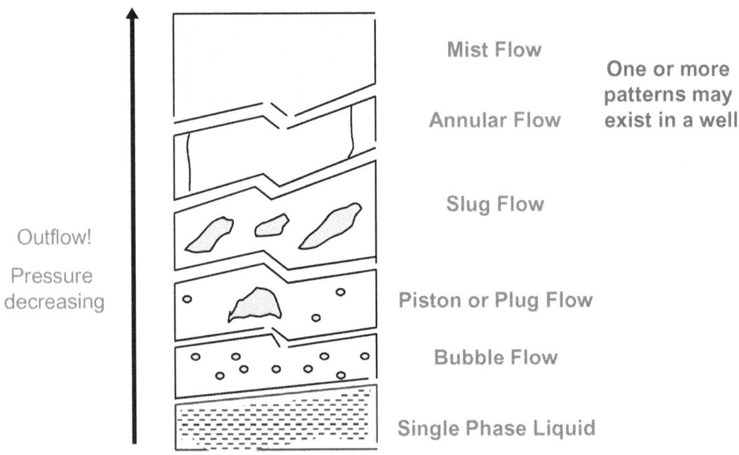

FIGURE 7.5 Multiphase Pattern in Production Tubing.

1. Dispersed bubble
2. Plug flow
3. Churn flow (equivalent of slug flow)
4. Annular flow
5. Mist flow (evidenced by high rate gas flow through choke system and liquid drop back into well)

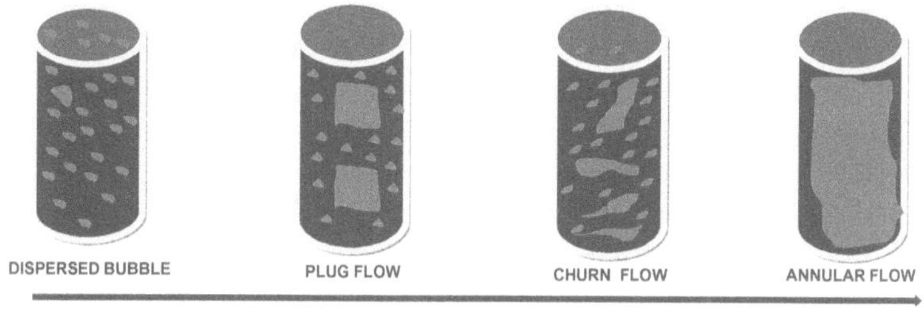

FIGURE 7.6 Multiphase Flow Patterns in Vertical Pipe.

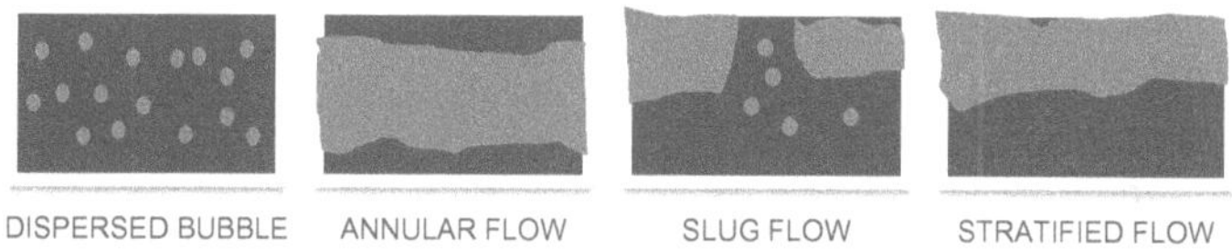

FIGURE 7.7 Multiphase Flow Patterns in Horizontal Pipes.

For oil-dominated systems the possible flow patterns are dispersed bubble and intermittent flow. For relatively low gas and liquid rates a stratified configuration occurs with the liquid flowing on the bottom and the gas flowing above it. It is important from the designer's point of view to be able to predict accurately which flow pattern will occur for given input flow rates, pipe size, and fluid properties. Only then can the appropriate flow model be adopted.

7.3 MULTIPHASE CORRELATIONS FOR FLOW IN WELL BORE AND PIPELINES

Modeling of production systems involves a large amount of calculating for multiphase flow in well bores and surface pipelines. Among the number of multiphase flow correlations available, none of them works well across the full range of conditions encountered in oil and gas fields.

7.3.1 Multiphase Flow Models[24–26]

In general terms, the pressure required to transport a specified volume of fluid from point A to point B consists of the following:

1. Frictional component
2. Elevation component
3. Pipe delivery pressure

The pressure drop along the flowline is represented by

$$\Delta P = \Delta P_f + \Delta P_{el} + \Delta P_{acc} \tag{7.1}$$

In a horizontal pipe arrangement, pressure loss due to elevation will be zero; therefore, total pressure loss is built by friction and acceleration only. As a rule of thumb, prediction of liquid holdup is less critical for pressure loss calculation in horizontal pipes, compared to vertical or inclined arrangements. But nevertheless the liquid holdup still needs to be calculated.

The pressure drop in a pipe due to friction can be represented by:

$$\Delta P = f_{tp} \frac{L}{D} \rho_m \frac{U_m^2}{2} \tag{7.2}$$

where:

f_{tp} = Two-phase friction factor
L = Pipeline length
D = Diameter
U_m = Mixture velocity
ρ = Slip mixture density

$$\rho_{slip} = \rho_L H_L + \rho_G H_G \tag{7.3}$$

Pressure drop due to elevation is given as

$$\Delta P_{el} = \rho_m g L \sin\theta \tag{7.4}$$

where:

L = Segment length
g = Acceleration due to gravity
θ = Angle of segment to horizontal
ρ_m = In-situ mixture density

The friction factor '*f*' depends on the Reynolds number 'Re' of the pipe flow, and the relative roughness *e*/*D* of the pipe wall:

$$f = f\left(\text{Re}, \frac{e}{D}\right) \tag{7.5}$$

For laminar flow ($\text{Re} < 2000$), 'f' can be estimated using

$$f = \frac{64}{\text{Re}} \tag{7.6}$$

The pressure difference between two points can be written with the Bernoulli equation:

$$\text{P}_1 - \text{P}_2 = \lambda \frac{\text{L}}{\text{d}} \frac{\rho \text{v}^2}{2} + \sum \zeta \frac{\rho \text{v}^2}{2} + (\rho_2 \text{gh}_2 - \rho_1 \text{gh}_1) + \frac{(\rho_2 \text{v}_2{}^2 - \rho_1 \text{v}_1{}^2)}{2} \tag{7.7}$$

On the right side of the equation there are four terms that represent friction caused by pipe: local friction, gravity, potential energy difference, and kinetic energy difference, respectively. The complex nature of multiphase flow in pipes has resulted in different methods being proposed for pressure drop. An example list of correlations is presented in Table 7.2. The pressure gradient equation for a multiphase system is generally developed based on Eq. (7.7), which can also be written as:

$$\frac{dP}{dL} = \rho_m \frac{g}{g_c} \sin\theta + \frac{f_{tp}\rho_f V_m^2}{2g_c d} + \frac{\rho_a V_m}{g_c}\frac{dV_m}{dL} \tag{7.8}$$

A number of variations of the above equation exist based on the results of analysis by different authors. One of the most popular correlations – the Beggs and Brill Model – is presented in here. Beggs and Brill proposed a general pressure gradient correlation for two-phase flow given by:

$$\frac{dP}{dL} = \frac{2f_{tp}\rho_{ns}V_m^2}{d_e} \tag{7.10}$$

where f_{tp} is two-phase friction factor, ρ_{ns} is no slip density, and V_m is mixture velocity

The equations to use depend on the prevailing multiphase flow pattern.

7.4 FLOW PATTERN CORRELATION

Examples of correlations suggested by Beggs and Brill[25] for the prediction of flow patterns in gas-liquid flow in pipes applicable for both horizontal and vertical pipes are as follows:

$$N_{Fr} = \frac{u_m^2}{gD} \tag{7.11}$$

$$\lambda_L = \frac{q_L}{q_L + q_G} \tag{7.12}$$

$$L_1 = 316\lambda_L^{0.302} \tag{7.13}$$

$$L_2 = 0.0009252\lambda_L^{-2.4684} \tag{7.14}$$

$$L_3 = 0.10\lambda_L^{-1.4516} \tag{7.15}$$

$$L_4 = 0.50\lambda_L^{-6.738} \tag{7.16}$$

The following relations will determine the flow patterns as suggested by Beggs and Brill.

For segregated (stratified) flow will exist with

$$\lambda_L < 0.01 \;\&\; N_{Fr} < L_1 \text{ OR } \lambda_L \geq 0.01 \;\&\; N_{Fr} < L_2 \tag{7.17}$$

For intermittent (slug) flow will exist with

$$0.01 \leq \lambda_L < 0.4 \;\&\; L_3 < N_{Fr} \leq L_1 \text{ OR } \lambda_L \geq 0.4 \;\&\; L_3 < N_{Fr} \leq L_4 \tag{7.18}$$

For bubble or dispersed bubble flow will exist if

$$\lambda_L < 0.4 \,\&\, N_{Fr} \geq L_1 \text{ OR } \lambda_L \geq 0.4 \,\&\, N_{Fr} \succ L_4 \tag{7.19}$$

Transition flow if

$$\lambda_L \geq 0.01 \,\&\, L_2 < N_{Fr} \leq L_3 \tag{7.20}$$

7.5 LIQUID HOLDUP

Most pressure-loss prediction correlations for two-phase flow in horizontal pipes as well as inclined surfaces require accurate prediction of two key parameters such as liquid holdup and two-phase friction factors. The reliability of these two parameters largely determines the accuracy of the pressure-drop prediction correlation.

Beggs and Brill defined liquid holdup in terms of flow patterns. The expressions for each of the flow patterns are as presented as follows.

For segregated flow

$$H_L = \frac{0.98\lambda^{0.4846}}{N_{FR}^{0.0868}} \tag{7.21}$$

For intermittent flow

$$H_L = \frac{0.845\lambda^{0.5351}}{N_{FR}^{0.0173}} \tag{7.22}$$

For distributed flow

$$H_L = \frac{1.065\lambda^{0.5824}}{N_{FR}^{0.0609}} \tag{7.23}$$

where the Froude number, input liquid content and mixture velocity are

$$N_{FR} = \frac{v_m^2}{gd} \tag{7.24}$$

$$\lambda = \frac{q_l}{(q_l + q_g)} \tag{7.25}$$

$$v_m = \frac{(q_l + q_g)}{A_p} \tag{7.26}$$

The following is an example procedure used to calculate the liquid holdup and pressure drop:

(1) Calculate total flux rate

$$U_M = U_{SL} + U_{SG} \tag{7.27}$$

(2) Calculate no-slip holdup

$$\lambda_{ns} = \frac{U_{SL}}{U_{SL} + U_{SG}} \tag{7.28}$$

(3) Calculate the Froude number, N_{FR}

$$N_{FR} = \frac{{U_M}^2}{gd} \tag{7.29}$$

(4) Calculate the liquid velocity number

$$N_{Lv} = U_{SL}\left(\frac{\rho_L}{g\sigma_L}\right)^{0.25} \tag{7.30}$$

(5) To determine the flow pattern that would exist if the flow was horizontal, calculate the correlating parameters, L_1, L_2, L_3, and L_4:

$$L_1 = 316\,\lambda_{ns}^{0.302};\ L2 = 0.0009252\,\lambda_{ns}^{-2.4684};\ L_3 = 0.10\,\lambda_{ns}^{-1.4516};\ L_4 = 0.5\,\lambda_{ns}^{-6.738} \tag{7.31}$$

(6) Determine flow pattern using the following limits:
Segregated: $\lambda_{ns} < 0.01\ and\ N_{FR} < L_1\ or\ \lambda_{ns} \geq 0.01\ and\ N_{FR} < L_2$
Transition: $\lambda_{ns} \geq 0.01\ and\ L_2 < N_{FR} \leq L_2$
Intermittent: $0.01 \leq \lambda_{ns} < 0.4\ and\ L_3 < N_{FR} \leq L_1\ or\ \lambda_{ns} \geq 0.4\ and\ L_3 < N_{FR} \leq L_4$
Distributed or dispersed bubble: $\lambda_{ns} < 0.4\ and\ N_{FR} \geq L_1\ or\ \lambda_{ns} \geq 0.4\ and\ N_{FR} > L_2$

(7) Calculate the horizontal holdup λ_0

$$\lambda_o = \frac{a\,\lambda_{ns}^b}{{N_{FR}}^c} \tag{7.32}$$

where a, b, and c are determined for each flow pattern as shown in Table 7.3.

(8) Calculate the inclination correction factor coefficient.

$$C = (1 - \lambda_{ns})\ \ln\left(d\,\lambda_{ns}^e\,{N_{Lv}}^f\,{N_{FR}}^g\right)$$

where d, e, f, and g are determined for each flow condition as given in Table 7.4.

(9) Calculate the liquid holdup inclination correction factor

$$\psi = 1 + c\left(\sin(1.8\theta) - 0.333\sin^3(1.8\theta)\right) \tag{7.33}$$

where θ is the deviation from the horizontal axis.

(10) Calculate the liquid holdup

$$\lambda = \lambda_o\,\psi \tag{7.34}$$

TABLE 7.2 Example List of Multiphase Flow Correlations[24]

Multiphase Correlation for Vertical Flow	Comments
Poettmann and Carpenter (1952)	Slippage between the phases and the flow regime not taken into account.
Baxendell and Thomas (1961)	Slippage between the phases and the flow regime not taken into account.
Fancher and Brown (1963)	Slippage between the phases and the flow regime not taken into account.
Hagedorn and Brown (1965)	Slippage between phases considered but flow pattern not taken into account.
Gray (1978)	Slippage between phases considered but flow pattern not taken into account.
Asheim (1986)	Slippage between phases considered but flow pattern not taken into account.
Duns and Ros (1963)	Slippage between phases, and the flow pattern considered. May not be applicable for well with water cut
Orkiszewski (1967)	Slippage between phases, and the flow pattern considered.
Aziz et al (1972)	Slippage between phases, and the flow pattern considered.
Beggs and Brill (1973)	Slippage between phases, and the flow pattern considered. Pipe angle other than vertical upward taken into account
Mukherjee and Brill (1985)	Slippage between phases, and the flow pattern considered. Pipe angle other than vertical upward taken into account.

TABLE 7.3 Beggs and Brill, *a*, *b*, *c* Constants[25]

Flow Pattern	*a*	*b*	c
Segregated	**0.98**	**0.4846**	**0.0868**
Intermittent	**0.845**	**0.5351**	**0.0173**
Distributed	**1.065**	**0.5824**	**0.0609**

TABLE 7.4 Beggs and Brill, *d, e, f, g* Constants[25]

Flow Pattern	*d*	e	*f*	*g*
Segregated uphill	**0.011**	**−3.768**	**3.539**	**−1.614**
Intermittent uphill	**2.96**	**0.305**	**−0.4473**	**0.0978**
Distributed uphill		**No** Correction C=O		

(11) Apply the Palmer correction factor:

$\lambda = 0.918 * \lambda$ for uphill flow
$\lambda = 0.541 . \lambda$ for downhill flow

(12) When flow is in transition pattern, take the average as follows:

$$\lambda = a\,\lambda_1 + (1-a)\,\lambda_2;\quad a = \frac{L_3 - N_{FR}}{L_3 - L_2} \tag{7.35}$$

where λ_1 is the liquid holdup calculated assuming flow is segregated and λ_2 is the liquid holdup assuming the flow is intermittent.

(13) Calculate the frictional factor ratio

$$\frac{f_{tp}}{f_{ns}} = e^S \tag{7.36}$$

where:

$$S = \frac{\ln(y)}{-0.0523 + 3.182\ln(y) - 0.8725[\ln(y)]^2 + 0.01853[\ln(y)]^4} \tag{7.37}$$

$$y = \frac{\lambda_{ns}}{\lambda^2} \tag{7.38}$$

(14) Calculate the frictional pressure gradient

$$(N_{\mathrm{Re}})_{ns} = \frac{\rho_{ns}\,U_M\,D}{\mu_{ns}} \tag{7.39}$$

Use this no-slip Reynolds number to calculate the no-slip friction factor, f_{ns}', using Moody's diagram, then convert it into the Fanning friction factor, $f_{ns} = f_{ns}' / 4$. The two-phase friction factor will be:

$$f_{tp} = f_{ns}\frac{f_{tp}}{f_{ns}} \tag{7.40}$$

TABLE 7.5 Empirical Constant for Two-Phase Critical Flow Correlations

Correlation	Reference	A	B	C
Gilbert (1954)	[108]	10.0	0.546	1.89
Ros (1960)	[111]	17.4	0.5	2.0
Baxendell (1967]	[109]	9.56	0.546	1.93
Achong (1961)	[110]	3.82	0.65	1.88

The friction pressure gradient is:

$$\left[\frac{dp}{dx}\right]_f = \frac{2f_{tp}\,\rho_{ns}\,U_M{}^2}{D} \tag{7.41}$$

7.6 MULTIPHASE FLOW THROUGH CHOKE[24]

Choke flow generally operates in 'critical' or 'sonic' flow conditions (i.e., the velocity of the fluids through the choke reaches a level identical to the velocity of sound), with the aim to restrict flow to the desired rate. Flow then becomes independent of downstream disturbance of pressure, temperature, or density because the disturbance cannot travel in the upstream direction.

The rate of multiphase flow through a choke and the upstream pressure are correlated by the following relationship:

$$P_I = \frac{Aq_L\,R_P^B}{d^C} \tag{7.42}$$

where:

P_I = Upstream pressure, which is also known as the wellhead pressure
q_L = Liquid production rate
R_p = Producing gas/liquid ratio
d = Choke diameter
A, *B*, *C* = Empirical constants given in Table 7.5.

7.7 MULTIPHASE SOLIDS TRANSPORT IN PIPES[27]

The solid transport mechanism in multiphase flow in pipelines is dependent on several parameters, the most important of which are the carrier fluid flow velocity and solid particle size. These two parameters also determine the flow regime, which exists when transportation of solid particles takes place. The key objective, however, is to keep the solid particles in suspension and/or rolling along the bottom of the pipe to prevent sand-bed formation. In transporting unprocessed oil and gas reservoir fluids, it is important to avoid settling of the solids entrained in the body of multiphase fluids. This can be done by keeping the multiphase reservoir fluid velocity in the pipelines above certain levels – referred to as minimum transport velocity (MTV) in this case. The MTV depends primarily on the type and size of the entrained solid. If solids settle, the area of the pipe available for flow will be reduced and the fluid velocity may tend to increase initially until such a stage where settled solids completely block the flow part.

There are systems of governing equations that have been developed for solid transport velocity based upon behaviour of water-sand and water-oil-gas-sand multiphase flow in pipes. The mathematical model involves balance equations deduced from mass and momentum conservation laws, constitutive models and forces due to drag force, gravitational force, buoyancy force, friction force, particle-liquid turbulent interaction force, particle-particle interaction force, and particle-pipe-wall interaction force.

In addition to the conservation laws for mass, energy, and momentum, there are additional laws that govern the rate at which these quantities are transported from one region to another in a continuous medium. These are called phenomenological laws because they are based upon observable phenomena and logic but they cannot be derived by more fundamental principles. These rate or transport models can be written for all conserved quantities (mass, energy, momentum, electric charge, etc.) and can be expressed in the general form as (Darby 2001)

$$\text{Rate of transport} = \text{Driving force/Resistance} \tag{7.43}$$

The ability to predict the behaviour of solid-liquid-gas flows is vital for the successful design and determination of optimum operating conditions of the production string and pipelines. The dynamics of these types of systems can be investigated through physical experiments or through numerical simulations.

7.7.1 Mechanism of Particle Dynamics

Generally, whenever there is relative motion between a solid particle and a flowing fluid, the solid particle will experience drag and lift forces from the surrounding fluid. Accurate predictions of these forces are crucial to MTV models and serve as input parameters.

Several solids transport models have been proposed in the literature for prediction of solid transport in pipeline. The major forces acting on a solid particle flowing in any pipe orientations are shown in Figure 7.8. For horizontal pipe, the dominant forces are lift (CL), drag (CD), gravity (FG), and buoyancy (FB) forces. The gravity and buoyancy forces referred to as static forces and can be expressed as

$$F_G = \frac{\pi d_p^3}{6} \rho_p g \tag{7.44}$$

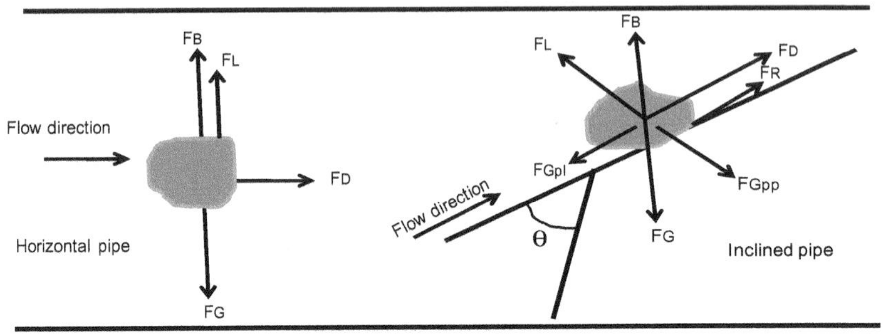

FIGURE 7.8 Forces Acting on a Solid Particle in Horizontal and Inclined Pipe Surfaces.

$$F_B = \frac{\pi d_p^3}{6} \rho_f g \tag{7.45}$$

The lift and drag forces are referred to as hydrodynamic forces and can be expressed as:

$$F_L = 0.5 C_L \rho_f V^2 A \tag{7.46}$$

$$F_D = 0.5 C_D \rho_f V^2 A \tag{7.47}$$

The moments acting on the particle due to the lift force, drag force, and buoyancy force tend to initiate movement, while the moment created by gravity tends to prevent the particle from moving. To initiate movement of the particle, the moments of forces (FB+FL+FD) that tend to initiate this movement must exceed the moments of the force (FG) that tend to prevent it. At the same time, the bed particle can also be lifted up if the sum of the forces in the upward direction is greater than the one in the downward direction. These can be expressed as

$$F_B + F_L + F_D > F_G \tag{7.48}$$

$$F_B + F_L > F_G - F_D, \text{ for rolling} \tag{7.49}$$

$$F_B + F_L > F_G, \text{ for suspension} \tag{7.50}$$

Friction forces are assumed to be zero at the beginning of particle movement.

In an inclined pipe surface, the forces acting on a particle being transported upward are predominantly the lift force (FL), drag force (FD), gravity force (FG), and friction force (FR) expressed as follows:

For gravity forces

$$F_{Gpl} = F_G \cos(\theta), \text{ parallel to the pipe axis} \tag{7.51}$$

$$F_{Gpp} = F_G \sin(\theta), \text{ perpendicular to the pipe axis} \tag{7.52}$$

For friction force the model can be expressed as

$$F_R = [F_G \sin(\theta) - F_L] f_s \quad (7.53)$$

$$F_R = \lfloor F_{Gpp} - F_L \rfloor \quad (7.54)$$

where *FGpl* is the resolved gravitational force parallel to the pipe axis, *FGpp* is the resolved gravitational force perpendicular to the pipe axis, and *fs* is the friction coefficient between the sand particle and the pipe-wall.

7.7.2 Concept of Minimum Transport Velocity

The sand particle transport driving forces are somewhat complicated in a transient multiphase flow environment. Multiphase fluid flow in a pipeline/tieback is a transient phenomenon. The flow of multiphase oil and gas production in the long tieback is accompanied by pressure drop, with the multiphase pattern generally changing as a result, from dispersed bubble through to slug, plug, annular, and stratified flow patterns depending on liquid-liquid-gas flow velocities, and pipe angle among other factors. Solids entrainment is subjected to different driving forces given different flow patterns as they are transported through the pipeline (see Figure 7.9).

For practical purposes and to simplify the complex phenomena, the concept of minimum transport velocity (MTV) best describes the physics of the transient solids transport mechanism The underlying principle of MTV is that solids in subsea tiebacks/pipelines or deep wells will be transported as long as they are upwardly mobile whether by rolling/sliding along the low side-wall of a pipeline or in heterogeneous suspension. The concept assumes that there

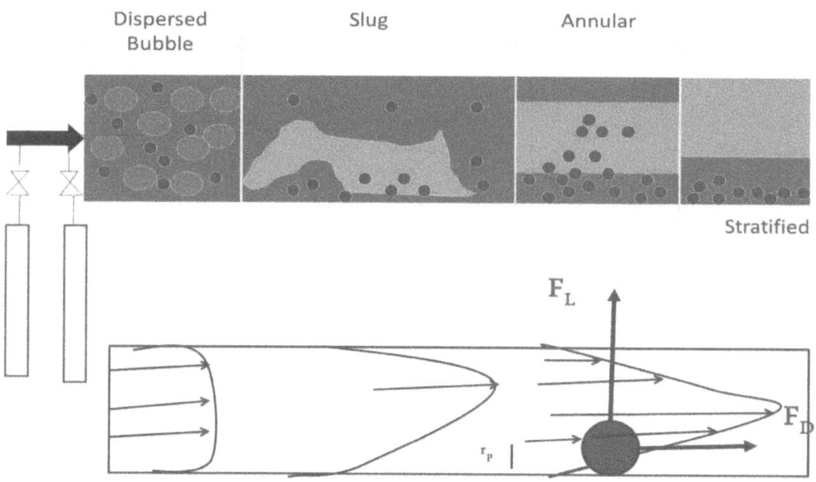

FIGURE 7.9 Multiphase Fluid and Solids Transport Phenomena Showing Trend of Flow Velocity Profile.

will as usual be a velocity distribution in the pipeline and the fluid point velocity acting on a solid particle on the low side-wall of the pipe (Figure 7.4) needs to be greater than the minimum transport velocity for the solid particle to be upwardly mobile.

Thus, for the average fluid velocity below the MTV for rolling, a stationary bed will result. When the velocity is below the MTV for suspension, this will result in solids sliding along the pipe-wall, which may also result in a stationary bed as the pressure drops along the pipeline, causing further reduction in the particle drag forces. However, the key focus here was integrating the velocity profile models for different flow patterns in the overall development of the solid transport models, which formed the basis of the MTV predictive models for suspension and rolling. The velocity profiles models developed for different flow patterns were discussed in Chapter 3.

7.8 MINIMUM TRANSPORT VELOCITY MODELS

The coupled equations for MTV rolling and suspension in horizontal, inclined, and vertical pipes can be expressed as:

For solid particle rolling

$$V_m = \left[\frac{a * g d_p \left(\frac{\rho_p}{\rho_f} - 1\right)(\cos\theta + f_s \sin\theta)}{(C_D + f_s C_L)}\right]^b \tag{7.55}$$

where fs can be expressed as

$$f_s = \frac{\sin\theta'_c}{\cos\theta'_c} \tag{7.56}$$

$$\theta'_c = 55 * \frac{\pi}{180} \tag{7.57}$$

For solid particle in suspension

$$V_m = a * \left[\frac{g d_p}{C_L \rho_f} * \left(\rho_p - \rho_f\right)\sin\theta\right]^b \left[\frac{D\rho_f}{\mu_l}\right]^c \tag{7.58}$$

For solid particle transport in vertical pipe

$$V_m = a * \left[\frac{g d_p \left(\rho_p - \rho_f\right)}{C_D \rho_f}\right]^b \tag{7.59}$$

where C_D and C_L are hydraulic drag and lift coefficients determined from deposit velocity and MTV rolling experiments respectively, ρ_P and ρ_f are solid particle density and fluid density, d_P is the particle diameter, a, b

and c are empirical constants as determined from the experimental data or numerical simulation results.

The magnitude of drag and lift coefficients are primarily dictated by the particle Reynolds number (Rep), defined as

$$R_{ep} = \frac{\rho_f V_f d_p}{\mu_f} \tag{7.60}$$

where:

R_{ep} = Particle Reynolds number
V_f = Fluid velocity, m/s
d_p = Particle diameter, m
μ_f = Fluid viscosity, cp

The minimum transport velocity models for suspension (Eq. 7.61) and rolling (Eq. 8) can be represented as:

For suspension

$$V_m = A * \left[\frac{g d_p}{C_L \rho_L} * (\rho_P - \rho_L)\sin\theta\right]^B \left[\frac{D\rho_L}{\mu_L}\right]^C \tag{7.61}$$

For rolling

$$V_m = \left[\frac{A * d_p\left[\frac{\rho_P}{\rho_f} - 1\right] g * [\cos\theta + f_s * \sin\theta]}{[C_D + f_s C_L]}\right]^B \tag{7.62}$$

For vertical pipe

$$V_m = A * \left[\frac{g d_p\left(\rho_p - \rho_f\right)}{C_D \rho_f}\right]^B \tag{7.63}$$

where C_D and C_L are drag and lift coefficients, ρ_P and ρ_f are particle density and fluid density, and d_P is the particle size.

a, b and *c* are defined as constants.

$$C_D = \left[\frac{a}{\mathrm{Re}_p^b}\right]; C_L = \left[\frac{c}{\mathrm{Re}_p^d}\right] \tag{7.64}$$

$$\mathrm{Re}_p = \left[\frac{\rho_f v_p d_p}{\mu_f}\right] \tag{7.65}$$

where $\mathrm{Re_p}$ is the particle Reynolds number and a, b, and c are empirical constants that can be determined from the experimental data or numerical simulation results.

There are a number of software packages for analyzing particle transport in multiphase environment. A new software is being developed by Intelligent Flow Solutions, Ltd. (http://www.intelligent-flow.com[7]) for real-time solids transport analysis in multiphase pipelines, especially subsea tiebacks.

7.8.1 MTV Prediction Methodology and Condition Monitoring with SMARTMULTIPHASE[7]

1. For a given tubing size and angle, production rate and fluid PVT properties the average fluid velocity is calculated.
2. This is followed by the definition and tracking of the prevailing multiphase flow pattern and corresponding velocity profile.
3. The actual velocity profile can be computed analytically for laminar and turbulent flow as given by Eqs. 4 and 5

$$V_R = \frac{f}{8} * R_e * V * \left[1 - \left(\frac{r}{R}\right)^2\right] \tag{7.66}$$

If $f = \frac{16}{R_e}$ for laminar flow, then

$$V_R = 2V * \left[1 - \left(\frac{r}{R}\right)^2\right] \tag{7.67}$$

where:

V_R = Velocity of fluid particle at a particular point in the pipe cross-section

V = Average velocity of the fluid s in the pipeline

r = Distance from the pipeline centre to any point in the flow field

R = Radius of the pipeline

R_e = Fluid Reynolds number, which defines the fluid flow regime whether laminar or turbulent flow

f = Fluid flow friction factor, which is a function of the pipe roughness, fluid flow regime and type of fluid

Based on the above, the velocity profile model is developed for multiphase flow presented in Eq. 6 below. The model is strongly dependent on Reynolds number and pipe friction factor.

$$V_R = A * f * R_e^B * \left[1 - \left(\frac{r}{R}\right)^2\right]^C \tag{7.68}$$

where, f is the friction factor and Re is the Reynolds number. The constants *A*, *B*, and *C* are proprietary empirical constants.

4. For the produced formation sand the particle Reynolds number and drag/lift coefficients are established

5. Thereafter the corresponding MTV for rolling and suspension are evaluated
6. If the average fluid velocity is less than MTV for rolling, no transport takes place
7. For velocity between rolling and suspension MTV, particle transport
8. Prediction of stationary bed is accompanied by bed height evaluation.

7.8.2 Introduction to Background Sand Transport Mechanism

Flow assurance across the composite production system from reservoir through the well bore to the topside facilities is one of the major concerns in multiphase sand transport in production operations. It is therefore critical that there is an accurate estimate of sand minimum transport velocity (MTV). In the development of predictive models for solid transport the objective is to propose models that will apply to all flow patterns at any inclination.

Consequently, the accuracy of solid transport models depends on how well the hydrodynamics is described and predicted with sufficient reliability. Therefore, to establish accurate predictive model(s) for minimum transport velocity, it is appropriate to understand the mechanism of interaction of fluid flow and sand particle movements. The best way to achieve this is to determine the velocity profile in a multiphase flow especially for each of the flow patterns.

7.8.3 Solid Transport Mechanism and Application of Minimum Transport Velocity

The solid transport mechanism is dependent on several parameters of which the most important are the flow velocity and particle size. These two parameters also determine the flow regime that exists when transportation of solid particles takes place. However, the key objective is to keep the solid particles in suspension and/or rolling along the bottom of the pipe to prevent sand-bed formation.

For flow of solid, liquid, and gas mixtures in production tubing and the liquid and solid phases may distribute in a number of configurations. The characterised three regimes of sand motion are characterised as:

- Flow with a stationary bed
- Flow with a moving bed and saltation (with or without suspension)
- Heterogeneous mixture with all solids in suspension

The solid particles will often be deposited at the bottom of the pipe if the velocity of the liquid and/or gas is very low and will form a stationary bed. The bed becomes stationary when the sum of the driving force acting on the bed is lower than the sum of forces opposing the bed motion. The sand grains are either rolling or saltating along the bottom of the pipe when there is increased velocity to keep the solid moving. The transition between a heterogeneous suspension and a sliding bed is often dependent on whether the velocity is decreasing or increasing. The particle transport driving forces are further complicated in a transient multiphase flow environment. Multiphase fluid flow is a transient phenomenon. The flow of multiphase production in the long production tubing will be accompanied by pressure drop, with the multiphase pattern generally changing as a result, from dispersed bubble through to slug, plug, annular, and stratified flow patterns depending on liquid-liquid-gas void fractions and pipe angle among other factors. Solids entrainment in the different flow patterns as they are transported will therefore be subjected to different driving forces. Of greater importance, therefore, is the need to know the multiphase pattern change and if, when, and where solids will settle in the pipeline. The critical questions, especially with respect to tiebacks, are:

- How much sand is produced into the production tubing?
- What are the prevailing multiphase patterns, the transition zones, and impacts on the solid-fluid interaction?
- Whether the solids will settle given the inherent operational conditions (probably yes).
- When the solid settles, how much will it be and what will be the height of sand in the pipe?
- Where will the solids settle?

Among other critical issues, these pose a serious threat to flow assurance. It is desirable to avoid the formation of a stationary deposit while transporting the entrained solids in multiphase fluids, which may cause complete/partial blockage of the pipe, thus reducing its efficiency.

Detailed quantitative risk assessment of solids transport in deepwater producing well completion tubing requires accurate analysis with the dedicated proprietary SMARTMULTIPHASE software developed by Intelligent Flow Solutions. For practical purposes and to simplify the complex phenomena, the concept of MTV transport mechanism has been adopted. The underlying principle of the MTV concept is that solids in the tubing will be transported as long as they are upwardly mobile, whether by rolling/sliding along the low side-wall of a pipeline or in heterogeneous suspension. Therefore, there exists a minimum or critical velocity for each of these transport mechanisms to occur. These are referred to as the MTV for rolling and MTV for suspension. Thus, for average fluid velocity below the MTV for rolling, a stationary bed will result. For velocity below the MTV for suspension, this will result in the solids sliding along the pipe-wall that may eventually result in a stationary bed as the pressure drops along the pipeline, causing further reduction in the particle drag forces.

In view of this initial sand management concept, it is important to provide input into the subsea and topside design in terms of possible quantity, type based on particle size distribution (PSD), and frequency of sand produced and carried up through the well bore into the subsea and topside facilities by making a quantitative prediction of sand production over the well/field life. This quantification is also important for the well design as, in the case of the multizone wells, any sand produced from the upper zone will tend to fill up the finite volume in the annulus behind the tubing and above the packer below and eventually plug up the perforations and/or the ICV (Inflow Control Valve) and pose a flow assurance problem.

The case study results presented here require evaluating well log and drill stem tests, rock strength, and mineralogy data and assessing various failure modes over well life to quantify sand production.

In order for the facilities and subsea network to be designed appropriately, some quantification on the amount and type of sand likely to be produced during the well/field life and the amount and type likely to be carried up the well bore, through the subsea network and up to the topsides, needs to be provided. More details on this topic are given in Chapter 8.

Chapter 8

Risk Assessment Criteria for Effective Sand Management

Chapter Outline

8.1 INTRODUCTION

The major challenges for asset teams in the development of oil/gas fields in deepwater environments, especially in relation to total sand management, are in how to:

- Provide inputs into the field/subsea and topsides design in terms of possible quantity and particle size distribution of sand and frequency of sand produced and transported through the wellbore into the subsea and topside facilities
- Optimise a well design that is fit-for-purpose
- Maximise the individual well's production and field development performance
- Effectively manage the sand and multiphase fluid production as well as the integrity of the facilities
- Deploy an appropriate sand management strategy that is fit-for-purpose including critical evaluation of the appropriate sand control method – surface control vs subsurface
- Minimise the impact of sand production on the well and subsea production facilities
- Minimise operational cost and non-productive time and guarantee flow assurance end-to-end across the entire composite production system from reservoir through the wellbore to the topside and flow lines
- Manage the waste disposal, including the environmental impact assessment arising therefrom

Developments in Petroleum Science, Vol. 63. http://dx.doi.org/10.1016/B978-0-444-62637-0.00008-7

Achieving these objectives is the driver for the unique, integrated geological and engineering solutions strategy presented in this chapter. The following subsections presents a step-by-step total sand management solution strategy that can enable managers of subsea installations and complex fields to improve intervention and production efficiency through continuous process optimisation, reducing non-productive time and also guaranteeing flow assurance that can reduce lifting costs per barrel.

The studies carried out over the years by the author and his research group[28] show that log-based data and corresponding over and above core data provide the best approach to detailed reservoir characterization defined by the reservoir quality index (RQI) and flow zone indicator (FZI) required for sand production rate prediction and transport analysis.

Recent field development studies by Intelligent Flow Solutions (http://www.intelligent-flow.com[2]) appear to also indicate that charging the reservoir with water injection can be an effective sand control method by extending the sand failure envelope and extensively prolonging the onset of sand production.

8.2 TOTAL SAND MANAGEMENT STRATEGY

The risk assessment strategy should focus on and provide comprehensive answers to the following:

- If and when will a well produce sand?
- What is the sand production rate prediction?
- Can we prevent or reduce the sand production without compromising fluid production?
- How do we manage a sand producer?
- What is the production optimization?
- What is grain size – pore size profiling?
- What is pore pressure/depletion profiling?

In addressing these issues it is important to develop integrated sand and flow assurance management solutions that, in addition to the sand quantification, cover:

- Fit-for-purpose sand control selection strategy
- Sand control completion design and process optimisation
- Development of a corporate integrated sand management strategy
- Process engineering design of a fit-for-purpose topside solids control system

The strategic approaches of importance include:

i. Initial detailed consultation within the asset/operation team to establish the goal/objectives and the range of data required and available
ii. Data compilation and verification

iii. G & G data analysis to establish the key pay sections and provide the necessary petrophysical and flow properties parameters required as input data for the engineering analysis.
iv. Detailed engineering data compilation
v. Detailed engineering studies to predict sand production rate/grain size distribution as well as model the corresponding sand control design and transport rate analysis in the wellbore and long subsea tieback

In fulfilling this strategy the datasets required in varying details are:

1. Core data including the poroperm profiles
2. Well test evaluation data
3. End of well report (EOWR)
4. Geomechanical data
5. Log data covering mostly the wireline log data generated
6. PVT data covering mostly the composition, fluid properties, and modular dynamic tester (MDT) data reservoir fluid pressure profile

The key data extraction and analysis required can be divided intos:

1. Geological and geophysics data extraction and analysis for use as key input parameters for the hydrocarbon reservoir characterisation data required for the engineering risk assessment
2. Conventional engineering data extraction and analysis derived for key fluid properties, well geometry, and operating conditions

The underlying premise for the detailed data analysis from each of the sources is to establish the consistency or otherwise of each of the derived datasets, first for QA/QC and secondly to identify the most accurate and representative data sets required for engineering prediction studies.

8.2.1 G & G Data

For the G & G studies, the following procedure is recommended:

1. Extract data from
   a. End of well report (EOWR) for field and well history, available interpreted data
   b. Core data obtained from sidewall core analysis for poroperm analysis and geomechanical studies. Core data covered includes:
      i. Core description and sedimentology
      ii. Core logs, core petrography, core photos
      iii. Routine core analysis and special core analysis including geomechanical studies
      iv. Whole core or side-wall core analysis
   c. Log data covering GR, neutron porosity, bulk density, resistivity, and special CMR+ log suite

2. Map out and review inventories and correlate data for completeness and synergy for all the planned wells and those in production in a field.
3. Interpret comprehensive log data to:
   a. Define lithologies
   b. Pick reservoir tops and bases
   c. Define hydrocarbon-bearing zones
   d. Carry out detailed sedimentological studies
   e. Determine
      i. Net to gross (N/G) of the sand levels in a field
      ii. Hydrocarbon content (HC), types, and contacts
      iii. Porosity and permeability
      iv. Water saturation
      v. Permeability profiles

   Detailed log analysis needs to be carried out using fit-for-purpose IP software with calibrated and validated models to generate both S_w (water saturation) and porosity curves (porosity derived from density logs) from which both parameters can be determined at depths of interest. Also required for display are, for example:
   - KTIM (TIM/Coates) permeability
   - KSDR (Schlumberger Doll Research) permeability
   - TCMR (total CMR porosity)
   - CMFF (free fluid porosity)
   - CMRP_3ms (bound fluid porosity) curves all from the CMR or NMR suite of logs
   - Conventional logs such as GR, resistivity and TNPH (thermal neutron porosity)
4. Use log-derived petrophysical data to compute the permeability profiles to define reservoir heterogeneities and corresponding pore-size and grain-size distributions based on the special reservoir quality index/flow zone indicator (RQI/FZI) concept.
5. Carry out detailed QA/QC on the poroperm data generated from different core analysis databases

8.2.2 Engineering Data Requirements

The key engineering risk assessment details essential to effective sand management and flow assurance are:

- Sand production rate prediction
- Control/management of the sand production without compromising fluid production
- Production optimisation

- Grain size – pore size profiling
- Pore pressure/depletion profiling

The most important reservoir parameters that drive sand production rate prediction are porosity, permeability, and the geomechanical properties such as UCS (unconfined compressive strength) and TWC (thick-wall cylinder).

The G & G data provide useful input to the required analysis especially with respect to porosity data. However, permeability data is a unique entity. These can be sourced not just from core data but also from well test data as well as the reservoir quality index. The RQI/FZI concept can provide the most consistent permeability data generator at P10 and P90 levels. For the sand transport analysis, the formation sand grain size and produced sand volume are essential elements.

8.2.3 Reservoir Quality Index (RQI) Analysis

The reservoir quality index (RQI) concept is a unique and useful way to quantify the flow character of a reservoir through the use of the FZI (flow zone indicator). The FZI offers a relationship between petrophysical properties at a small scale, such as core plugs, and on a large scale, such as well-bore level. The FZI provides the representation of the flow zones based on the surface area and tortuosity. Based on this concept the permeability profile and the corresponding homogeneity or heterogeneity footprint for each sand level can be defined.[29,30]

The permeability profile can be computed from the following log data as follows:

$$k = [FZI * \{\phi/(1-\phi)\}/0.00314]^2 \quad (8.1)$$

where

$$FZI = M^*SUMTr^2 + N^*SUMTr + O \quad (8.2)$$

where $SUMTr = GR\_Tr + NPHI\_Tr + RHOZ\_Tr + LLD\_Tr$

$$The\ normalised\ GR\_Tr = A^*E-03\ GR^2 - B^*GR + C^*E + 00 \quad (8.3)$$

$$The\ Neutron\ Porosity,\ NPHI\_Tr = D^*NPHI^2 + E^*NPHI + \mathrm{F} \quad (8.4)$$

$$The\ Formation\ Density\ RHOZ\_Tr = G^*RHOZ^2 - H^*RHOZ + I \quad (8.5)$$

$$The\ Resistivity\ LLD\_Tr = J^*HLLD^2 - K^*LLD + L \quad (8.6)$$

A, B, C, D, E, F, G, H, I, J, K, L, M, N, and O are proprietary empirical constants upgraded from the original Amaefule model. The formation grain size [d_{50}] and formation pore size can be computed using relevant models.

A typical example of such equations are the following equations, modified here for proprietary reasons:

$$d_{50} = \frac{A(1-\phi)}{S}$$
$$k = \frac{\phi^3}{B(1-\phi)^2 S^2} \tag{8.7}$$
$$D_{pore} = \frac{D_{50}{}^* \phi}{C(1-\phi)}$$

where:

d_{50} = Average grain size
k = Absolute permeability
ϕ = Absolute porosity
S = Specific surface area
D_{pore} = Formation pore size
A, B and C = Empirical constants

For the engineering phase of the study the following typical reservoir, well, and fluid parameters are essential:

1. Oilrim data: k_V, k_H, ρ_o, μ_o, B_o, R_s, c_o, porosity; S_{oi}, S_{or}, x_e, h, y_e
2. Gas cap data: m, k, k_{rg}, μ_g, ρ_g, z_g
3. Aquifer data: a, k, k_{rw}, r_w, μ_w, c_w, B_w
4. Well data: r_w, L_w, $Q_{oinitial}$, min FBHP; P_i, D_p; t_{max}
5. Rock properties: plastic parameter, biot effect, poisson ratio, critical porosity' UCS, hollow cylinder, critical cavity pressure gradient
6. Perforation properties: perf phase, perf. penetration, perfradius, perf spf, perf. phasing
7. Initial stress conditions: total vertical stress change, total vertical depletion, initial water depth
8. Wellbore properties: deviation, azimuth, influence zone radius
9. Grain size [d_{50}]: obtained from RQI analysis
10. Sand concentration: (sand volume produced) obtained from sand rate
11. Fluid type and properties: from MDT and PVT data
12. Fluid production rate: from production (PLT) data

References

1. Oyeneyin MB. Total sand management solution for guaranteed flow assurance in subsea development. Society of Petroleum Engineers (SPE) Paper Number 17429, August; 2014.
2. International Energy Foundation(IEA). *IEA world energy outlook;* 2008.
3. Brooks D. Deepwater - challenges and opportunities. Paper presented at the Subsea and Deepwater Asia Conference, Singapore; 2008.
4. www.2hoffshore.com: SLOR Production System.
5. www.subseaUK.org.
6. Sawaryn SJ, Goodwin S, Deady A, Critchley C, Swanson B. The implementation of a Drilling and Completions Advanced Collaborative Environment – Taking Advantage of Change. SPE Paper No. 123801; 2009.
7. Walsh MP, Lake LW. *A generalized approach to primary hydrocarbon recovery.* Elsevier Publishers; 2003.
8. Oyeneyin MB, Maclellan G, Vijayakumar B, Hussain M, Bichan R, WierN. The Mare's Tail – the answer to produced water management in deepwater environment'? SPE Paper No 128609; 2009.
9. www.intelligent-flow.com.
10. www.sandmanagement.com.
11. Moriwawon B. *Real time prediction of sanding potential in clastic (sandstone) reservoirs.* PhD Thesis (RGU), September 2007.
12. Oluyemi GF, Oyeneyin MB. Analytical critical drawdown (CDD) failure model for real time sanding potential based on hoek and brown failure criterion. *J Petrol Gas Eng* 2010;**1**(2): 16–27. (Online) Available from: http://www.academicjournals.org/article/article1379512631_Oluyemi%20and%20Oyeneyin.pdf.
13. Perkins TK, Weingarten JS. Stability and failure of spherical cavities in unconsolidated sand and weakly consolidated rock. SPE 18244, In: Proc. 63rd annual technical conference and exhibition, Houston; Oct. 2–5, 1988.
14. Sanda JP, Kessler N, Wicquart E, Hannaford K, Deflandre JP. Use of porosity as a strength indicator for sand production evaluation. SPE Paper No 26454; 1993.
15. Raaen AM, Hoven KA, Joranson H, Fjaer E. FORMEL- a step forward in strength logging. SPE Paper No. 36533; 1996.
16. Vernik L, Bruno M, Bovberg C. Empirical relations between compressive strength and porosity of siliciclastic rocks. *Int J Geophys* 1993;**59**:420–7.
17. www.clampon.com.
18. www.emerson.com.
19. www.pmacsystems.com.
20. Procyk A, King GE. Use sand analysis to select screens for non-gravelpacked Completions. *World Oil* 1997, Nov.
21. www.weatherford.com.
22. www.slb.com.

23. Tiffin DL, King GE, Larese RE, Britt LK. New criteria for gravel and screen selection for sand control. SPE 39437; 1998.
24. Bello K. *Modelling multiphase solids transport velocity in long subsea tieback - numerical and experimental methods*. PhD Thesis, Robert Gordon University; 2013.
25. Beggs HD, Brill JP. A study of two-phase in inclined pipes. *J Petrol Tech* 1973;607–17.
26. Behnia M, Llic V. A simple correlation for estimation of multiphase pressure drop in an oil pipeline. *SPE Prod Eng* 1990;370–2.
27. Bello KO, Oyeneyin MB, Oluyemi GF. Minimum transport velocity models for suspended particles in multiphase flow revisited. SPE Paper No 147045; 2011.
28. www.rgu.ac.uk.
29. Oyeneyin MB, Macleod C, Oluyemi G, Onuku A. Intelligent sand management. SPE Paper No. 98818; 2005.
30. Oyeneyin MB. Total sand management solution for guaranteed flow assurance in subsea development. SPE Paper No 172429; 2014.

Epilogue

This book presents a step-by-step total sand management solution strategy that can enable managers of onshore/offshore including subsea installations and complex fields to improve development and production efficiency through continuous process optimisation, reducing non-productive time and also guaranteeing flow assurance that can reduce lifting costs per barrel.

Detailed risk assessment is an essential process to minimise the impact of sand production especially in a deepwater/subsea environment. This book attempts to provide some key guidelines that will help an oil/gas field sand management asset team in answering the key questions: Will each of the wells produce sand? When and how much sand will be produced? How can the sand production be prevented or reduced without compromising fluid production? How are the sand producers and the multiphase transportation managed while enhancing production?

Index

Note: Page numbers followed by "*f*" indicate figures, and "*t*" indicate tables.

N

O

P

Q

R

T

U

V

W

Y

FSC
www.fsc.org
MIX
Papier | Fördert gute Waldnutzung
FSC® C083411

Kolibri 360 GmbH
Ferdinand-Jühlke-Straße 7
99095 Erfurt, Deutschland
produktsicherheit@kolibri360.de